I0681450

"Truth is stranger than fiction, but it is because fiction is obliged to stick to possibilities; truth isn't."

Mark Twain

"Truth really is stranger than fiction, if only because it is true."

Halloway Bumpsteed, Jr.

ASSIGNMENT: DESTINY!

VOLUME I

by

J. Rutger Buck

COPYRIGHT 2012

FIRST PRINTING
December 2014
SECOND PRINTING
March 2016

All rights reserved. No part of this book may be reproduced in any form, except for the inclusion of brief quotations for review, without the written consent of the author.

This book is a work of fiction. All references to historical events, real people, or real locales are used fictitiously. Other names, characters, places, and incidents are products of the author's imagination, and any resemblance to actual events, locales, or persons, living or dead, is entirely coincidental.

MANUFACTURED IN THE UNITED STATES OF AMERICA
WARRIOR SPARROW PRESS
http://warriorsparrowpress.com

This book is respectfully dedicated to all veterans, but most especially to all the Lieutenant Blakeleys who ever had to endure a Leo Hooper, a Captain Lacksdale, or a Major Claproot.

J. Rutger Buck

UNIONTOWN
CITIZEN ADVERTISER HOME

Cenr. 1941 by News Syndicate Co. Inc. UNIONTOWN'S PICTURE NEWSPAPER Trade Mark Reg. U. S. Pat. Off.

Average net paid circulation
for February exceeded
Daily ------2,000
Sunday-----4,000

Vol. 26. No. 49   UNIONTOWN, PA. ★ AUGUST 22, 1942   36 Main +1 Comic + 24 Coloroto Pages

GERMAN TERRORISTS ATTACK UNIONTOWN!

STORY ON PAGE 5

PRESIDENT AT WARM SPRINGS

BRITAIN'S AIR FORCES POUND DUCE'S TROOPS

Forces Sent From The Italian Mainland Are In Attack

German Port of Hamburg Placed On 'Useless' List By RAF Bombs

CALL SOUNDED FOR DRAFTEES IN THIS AREA

Man Is Treated At City Hospital Following Attack

BRITAIN'S AIR FORCES POUND DUCE'S TROOPS

Forces Sent From The Italian Mainland Are In Attack

PETAIN OUSTS VICE PREMIER

DEFENSE HEADS

ASSIGNMENT: DESTINY!

VOLUME I:

The Inauspicious Beginnings

Of The Flungk Z-44 Mk I "Boomerang"

"The dim warehouse of history holds many strange tales, but few so strange as this. Who in their wildest imaginings would have thought it possible that one of the most secretive aeronautical programs of the Second World War would result from the chance encounter of a four-year-old child and a band of stone-age savages?"

From: *Whackos, Crackos, Sickos and Psychos--A Psychological Study of Fruitcakes and the Nut-cases Who Warped Them*, by Dr. Waldo von Heinkerblonker, M.S., PhD, L.S.M.F.T. Out of print.

PROLOGUE

A Nation Reeling on the Ropes . . . .

Somewhere in the South Pacific, Early Summer, 1942:

U.S. pilots across the South Pacific doggedly slug it out, pitting their inferior P-40B Warhawks and ex-lend-lease Bell P-400 Airacobras against the vastly superior Mitsubishi AN6M Zero-Sens of the Empire of the Rising Sun.

Caught with their pants down at Pearl Harbor on December 7, 1941, the U.S. Army Air Corps is, eight months later, still reeling on the ropes. Valiant airmen in the South Pacific are stretched to the limits of manpower and equipment. Flying aircraft that were obsolete before the war began, these brave soldiers strive, in vain, to stem the vicious onslaught of the Chrysanthemum tide.

Spurred by the overwhelming superiority of their fighters, the Imperial Army and Navy have pushed their advantage relentlessly. The Japanese aviators facing

Curtiss XP-40. First flown in October, 1938, the P-40 was a front-line fighter of the U.S. Army Air Corps when the U.S. was attacked at Pearl Harbor. Originally conceived as a ground attack aircraft, the lack of a two-stage supercharger rendered it ineffective at high altitudes. At lower altitudes, the P-40 could be out-turned by the Japanese Zero, although General Claire Chennault's American Volunteer Group in China achieved impressive results with the airplane, though this was due mainly to Chennault's grasp of aerial tactics. Clearly, something more was needed.

The dreaded Mitsubishi AN6M Zero, scourge of the Southern Pacific skies. Its deadly reputation was well earned.

the beleaguered allies are already seasoned veterans of campaigns in Manchuria and China.

The Mitsubishi AN6M Zero, in particular, has quickly achieved an almost mythical status. First introduced into the China theatre of war in 1940, its wild maneuverability soon became legendary. The Chinese pilots, who had received some forewarning intelligence about the new airplane, were reluctant to engage the superior Japanese aircraft. When the Zeros finally caught the Chinese in the air over Chungking, 12 Zeros shot down 27 Chinese aircraft in a matter of minutes. Unfortunately for the allied forces in the South Pacific, it's all been downhill from there.

In June, 1942, the unfortunate pilots of Marine squadron VMF-22, tasked with defending Midway Island, respond to an attacking force of Japanese bombers and fighters. The stalwart Marines are equipped with the Brewster F2-A3 Buffalo, an overweight, underpowered, undergunned, bureaucracy-created monstrosity whose only fame today is that it is consistently placed at the top of any "Worst Fighter Ever Produced" list. In the ensuing battle, the Zeros shoot down 15 of the 21 Buffaloes, plus two Grumman F4F Wildcats, prompting one of the surviving Buffalo pilots to say, "It is my belief that any commander who orders pilots out for combat in an F2A should consider the pilot as lost before leaving the ground."

Although the Army Air Corps is fairing a little better than the Navy and Marines, this is due largely to the fact that General Claire Chennault's American Volunteer Group in China had developed tactics to allow their outmoded Curtis P-40B's to successfully combat the Zero. The other mainstay of the Army Air Corps is the Bell P 39, known, less than affectionately perhaps, as the "Iron Dog." Its numbers are strengthened by the addition of an export version of that aircraft that the British had previously rejected as unfit for combat: the P-400, known, perhaps even less affectionately, as "A P-39 with a Zero on its tail."

Even the exceptional firepower of "The Iron Dog" proved no match for the Zero.

The Brewster F2-A3 Buffalo, considered by many to be a textbook example of the problems with the procurement practices of the U.S. armed forces just prior to World War II.

The Japanese are, all too frankly, kicking the Air Corps' butts. If the slide-rule boys back in the states can't pull a rabbit out of their hat soon, preferably a 450 mile-an-hour rabbit with six fifty-caliber machine guns and the turning radius of an East Tennessee beagle, it looks like the folks stateside will be eating rice-balls with Japanese military-issue chopsticks quicker than they can say "Aragoto, Yamamoto."

As if this weren't enough, the Germans are pressing their air superiority advantage in the European

The world's first operational jet to see combat, the Messerschmitt Me-262 had already made it's maiden flight in July of 1942. Plagued by development problems with its Jumo 004-B turbojet engines, the Me-262 would not see combat until April of 1944, too late to affect the outcome of the war. Nonetheless, the development of fighter types such as these by the Axis powers was justly feared by the U.S. War Department, prompting the search for an advanced air superiority fighter.

Theater. Although their primary front-line fighter, the Messerschmitt BF-109, was originally designed in 1935, it more than holds its own against the Allied fighters that it is pitted against. At the same time, vague rumors and reports of outlandish and futuristic German aircraft designs circulate through the halls of U. S. Intelligence, sending the War Department into a frenzied search for any concept, no matter how incredible, for an advanced air superiority fighter. For the Allies, these are indeed the darkest days of the Second World War. It is under these conditions that The U.S. War Department comes to a fateful decision: Search outside the traditional sources for a tactical answer to the overwhelming domination of Axis airpower.

EDITOR'S FOREWORD:
SORDID, INSIGNIFICANT, AND FORGOTTEN:
THE STORY OF ONE D.S.P. VENTURE

What you are about to read is the story of only a single project as it was developed by the Department of Special Projects. The path of that development was to be ultimately revealed to J. Rutger Buck by a bulging box of documents that he received in response to his request for the files listed on D.O.D. Record of Declassification Invoice #C76497Q shown on the page opposite.

The particular venture addressed herein is in itself so insignificant, that history has largely ignored it. The players are either deceased, or prefer anonymity, or both, in some cases. But by following the shrouded and often convoluted narrative relating to this one small project, the editors sincerely hope that the reader will gain an overview of the function of this highly secretive department of our government known as the Department of Special Projects.

The documents reviewed herein are many and varied. Some of them are composed primarily of the written word; others are illustrations, blueprints, and, in some cases, photographs. They have originated from diverse and sometimes unique sources. At times, one of these documents will verify another of the documents; at other times, two or more of the documents will baldly contradict each other.

We, the editors, have attempted to keep our commentary to a minimum, hopefully adding to the narrative only where it would serve to clarify, rather than obscure.

Individually, these items contained in D.O.D. Record of Declassification Invoice #C76497Q amount to little more than historical oddities. Yet, taken together, they disclose a web of barely believable events that stuns the credulity of even the most gullible researcher, not to mention the casual reader.

It is the editors' sole intent that, by the reader's reviewing of these documents, a larger truth will emerge from the tangled nest of lies, half-truths, ignorance, deceit, and bald fabrication that comprises the bulk of these materials. The editors hope that the reader will exercise a moderate amount of patience as, together, we collaborate in revealing these arcane secrets of the Department of Special Projects. A state of confusion should be considered by the reader to be a natural reaction to this labyrinth of manipulation. It certainly was a natural state for the editors. But again, have patience.

Ultimately, all will be revealed.

Sincerely:
The editors
Warrior Sparrow Press

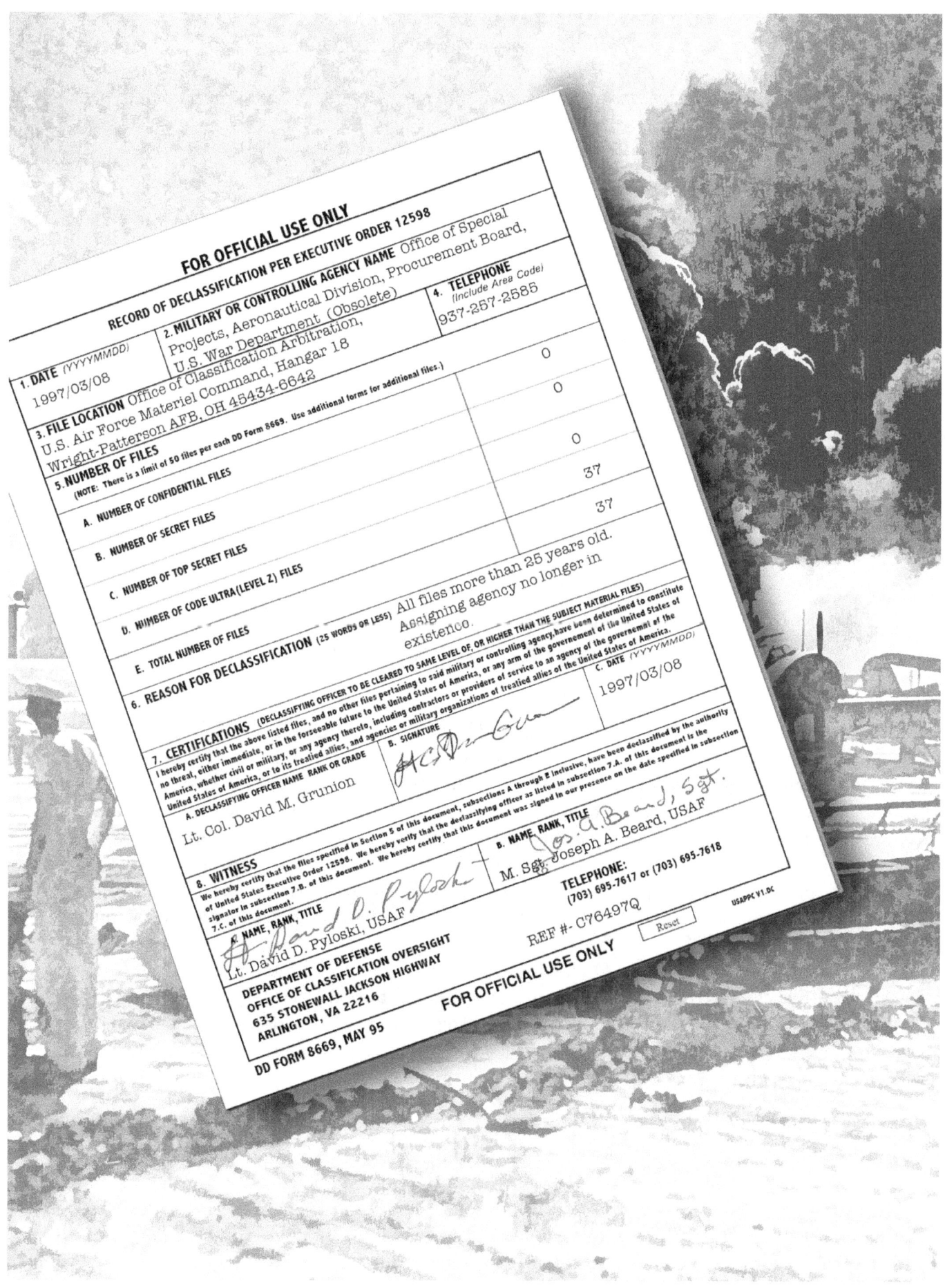

FOR OFFICIAL USE ONLY

RECORD OF DECLASSIFICATION PER EXECUTIVE ORDER 12598

1. DATE (YYYYMMDD)	2. MILITARY OR CONTROLLING AGENCY NAME Office of Special Projects, Aeronautical Division, Procurement Board, U.S. War Department (Obsolete)	4. TELEPHONE (Include Area Code)
1997/03/08	U.S. War Department (Obsolete)	937-257-2585

3. FILE LOCATION Office of Classification Arbitration, U.S. Air Force Materiel Command, Hangar 18 Wright-Patterson AFB, OH 45434-6642

5. NUMBER OF FILES
(NOTE: There is a limit of 50 files per each DD Form 8669. Use additional forms for additional files.)

A. NUMBER OF CONFIDENTIAL FILES	0
B. NUMBER OF SECRET FILES	0
C. NUMBER OF TOP SECRET FILES	0
D. NUMBER OF CODE ULTRA (LEVEL Z) FILES	37
E. TOTAL NUMBER OF FILES	37

6. REASON FOR DECLASSIFICATION (25 WORDS OR LESS) All files more than 25 years old. Assigning agency no longer in existence.

7. CERTIFICATIONS (DECLASSIFYING OFFICER TO BE CLEARED TO SAME LEVEL OF, OR HIGHER THAN THE SUBJECT MATERIAL FILES)
I hereby certify that the above listed files, and no other files pertaining to said military or controlling agency, have been determined to constitute no threat, either immediate, or in the forseeable future to the United States of America, or any arm of the governement of the United States of America, whether civil or military, or any agency thereto, including contractors or providers of service to an agency of the governement of the United States of America, or to its treatied allies, and agencies or military organizations of treatied allies of the United States of America.

A. DECLASSIFYING OFFICER NAME RANK OR GRADE	B. SIGNATURE	C. DATE (YYYYMMDD)
Lt. Col. David M. Grunion	*[signature]*	1997/03/08

8. WITNESS
We hereby certify that the files specified in Section 5 of this document, subsections A through E inclusive, have been declassified by the authority of United States Executive Order 12598. We hereby certify that the declassifying officer as listed in subsection 7.A. of this document is the signator in subsection 7.B. of this document. We hereby certify that this document was signed in our presence on the date specified in subsection 7.C. of this document.

A. NAME, RANK, TITLE	B. NAME, RANK, TITLE
[signature] Lt. David D. Pyloski, USAF	*[signature] Jos. A. Beard, Sgt.* M. Sgt. Joseph A. Beard, USAF
	TELEPHONE: (703) 695-7617 or (703) 695-7618

DEPARTMENT OF DEFENSE
OFFICE OF CLASSIFICATION OVERSIGHT
635 STONEWALL JACKSON HIGHWAY
ARLINGTON, VA 22216

REF #- C76497Q   [Reset]

FOR OFFICIAL USE ONLY

DD FORM 8669, MAY 95   USAPPC V1.00

(Editor's note: The following explanatory note accompanied the manuscript of "Assignment: Destiny!" when it was first received in our offices from Captain Buck. It is included here to clarify the manner in which Captain Buck came into possession of the documents upon which the manuscript is based.)

A CALL IN THE NIGHT

One night in late March, 1998, shortly before midnight, I received a telephone call from an anonymous source which, of course, must remain anonymous. Through the years, in both my professional life, and through my research contacts, I have met various members of the many clandestine organizations of the U.S. government. Some I consider friends; some remain professionally aloof, a few are outright hostile. Some, I think, use me as a conduit to the public for reasons of their own. Whatever their motivations for contact, (the one thing I am certain of, it is not my winning personality), it is not unusual for me to receive a late night call from one of these people on one of the few remaining public phones across the U.S. These calls may be for various reasons, but they all have one thing in common. Rarely does the caller identify himself. This was one of those calls.

As I could tell from my new caller ID service, this call originated from a number in an exchange that I recognized immediately. It was a Palmdale, California, number. As it so happened, I had spent some considerable time in Palmdale over the years. It is within easy driving distance of Edwards Air Force Base, and a few dozen other clandestine operations, both governmental and private, that you just didn't get into unless you were a member of a very exclusive club, and had the retina scans to prove it.

True to form, the caller would not identify himself. His voice sounded muffled, as though he was talking through a cloth obstruction of some sort. I envisioned him on a deserted road, maybe at an abandoned service station in the middle of the California high desert, with the chill wind blowing the white sand through the Joshua trees. At least I hope he was on a deserted road somewhere. The vision of someone at midnight in a crowded bar, talking into a pay-phone wrapped in a sock, would doubtless raise suspicions in the most casual observer.

The caller said it would be of interest to me to file a Freedom of Information request for a group of thirty-seven documents, and he began to rattle off a line of departments, bureaus, and boards. I groggily interrupted, telling him I needed to get a pen and paper. I grabbed at a notepad and pen that I keep on the night-stand for just such occasions. I keep a notepad there because, when I am abruptly roused from a deep sleep, I can often hold a fifteen minute conversation, then go back to sleep and promptly forget that I ever had the conversation. When I returned to the phone, though, there was only the buzz of a dead line.

I hung up the phone, crawled back into bed, and just as I reached for the light switch, the phone rang again. I answered the phone, and without preamble, the caller began to read off a list of information. I made my impatient caller stop, and repeat his list from the top, as I shakily copied it on the notepad in a kind of bogus shorthand that I hoped I could later decipher. Though still groggy, my interest was piqued, and I asked the caller what the files contained.

"I can't talk any longer," the caller gruffly interrupted me. "Get those files and see for yourself." Then he abruptly hung up. I shrugged, turned off the light, and went back to sleep.

The next morning, I stumbled down the stairs, ran some tepid tap water in a mug, dumped in some instant coffee, and tripped over the transom on my way out the front door. I was late for a breakfast meeting with an interviewee on another project, and it seemed like I spent all day catching up. When I finally got back to my bed around eleven o'clock, I saw the scribbling I had made the night before. I had forgotten all about it, but there it lay, on the night stand.

I was confused at first by the nearly illegible note, but suddenly the memory of the previous night's phone call came flooding back to me. Energized now by curiosity, I went downstairs to my computer. Little did I know that my real confusion was just beginning.

I TRY TO READ MY OWN HANDWRITING

I wanted to search for the various items I had written on the notepad, but they were barely decipherable. The one note that stood out for making the most sense, Wright Pat, was either the military way of presenting the name of someone called Pat Wright, or, more likely shorthand for Wright-Patterson Air Force Base, near Dayton, Ohio. I went on-line and found the home page for Wright-Pat, which I already knew to be home to the USAF Museum. Prominently displayed on the home page was the logo of the Air Force Materiel Command, which is also based there. This could have been the object of the reference scrawled on my notepad which seemed to read "Materal Comm."

I clicked on the link on the Wright-Pat home page for the website of "Air Force Materiel Command." Here I found the following mission statement: "Air Force Materiel Command: Equipping the Air Force with the best weapon systems through research, development, test and evaluation, acquisition management services and logistics support."

I tried searching for variations of "D.S.P.," "Aero Div," and "ProcrBd," all of which were scribblings on my notepad. Each time I got "0 results."

Still intrigued, but exhausted, I went to bed. The next morning, I called the main switchboard number for USAF Materiel Command. I was handed off or put on hold numerous times. I must have gotten just about every extension for USAF Material Command at every Air Force base in the U.S. Perhaps I even talked to my anonymous caller, I have no way of knowing. It was hard to know for sure without the sock. I did learn that, under the aegis of a new commanding officer, and in response to the newly instituted Executive Order 12598, a major housecleaning was in progress.

STRIKE ONE UP FOR WILLY

Executive Order 12598 related to the handling of classified documents, specifically the disposition of old or obsolescent documents. Signed into existence by President William J. Clinton in 1995, it contained one provision relating to classified material that directly affected the contents of this book. Before the order, if there was any doubt that material should be classified, the rule was: If in doubt, keep the material classified. Executive Order 12598 mandated the opposite: If in doubt, declassify the material. Hundreds of millions of pages of previously classified documents became subject to critical review regarding their classification status. These actions of review brought to light many previously unknown activities of various branches of our government, such as Psi experiments and remote viewing exercises, to mention just two of the most notably strange.

Because of the unique nature of the USAF Materiel Command, and its mission, their classified documents vault was bursting at the seams. Many of the files went back to the earliest days of the U.S. Army Air Corps. These files, many with top secret and above classification, were inherited from the agencies that preceded the USAF Materiel Command. Through the years, they remained classified, though the technologies they contained were long ago obsolete, and the major players long dead. It was my strong suspicion that they were still classified due more to inertia, rather than any need to protect the national interest.

The costs of maintaining these vaults of outmoded, yet still classified documents was enormous. Perhaps the new commandant saw Exec Order 12598 as a unique opportunity to get rid of a tremendous load of classified deadwood.

SPINNING MY WHEELS

All of this was very interesting, but it wasn't getting me any closer to the specific files that I was after. Or, more specifically, it wasn't getting me any closer to the files that Anonymous wanted me to access, for whatever reasons of his own. As I would later discover, Anonymous did me a great service, regardless of his motivations.

I looked at my note again. "Wardep" could mean War Depot, or War Deputy, or even Ward Epicenter, but these made absolutely no sense at all. Then again, it might refer to the U.S. War Department. This was the precursor to the U.S. Department of Defense, which was created when Harry S. Truman signed the National Defense Act into law in 1947. If this was the case, then these files would be ancient history. The U.S. War Department was an organization that had ceased to exist 50 years ago. For that reason alone, it was highly conceivable that classified files relating to that agency might indeed be considered obsolete, and therefore subject to purging. I just couldn't imagine that these files might contain anything of the remotest interest now. More specifically, how could they contain information so explosive that it would justify my surreptitious friend's midnight call?

I was getting nowhere fast. I was going around in circles with the USAF Materiel Command. That they were huge circles involving large numbers of evasive people was beside the point. Still, I was certain that this was the agency my tip had referred me to. This seemed especially possible for the reason that they were currently purging millions of pages of classified files from their vaults. But this only meant that finding the right documents would be a true needle in the haystack adventure.

TIME FOR AN END RUN

I decided to try another tack. I called the DOD Office of Classification Oversight. This is a tiny little organ of the Department of Defense behemoth that few people, even those with decades of experience in dealing with the Federal juggernaut, know about. Unless you maintained classified documents, or, like me, occasionally conducted research associated with classified documents, you would never hear of it in a million years.

The DOD Office of Classification Oversight is a sort of a clearinghouse for classified documents. It is nestled into a nondescript basement of a nondescript building in a nondescript town on the beltway of Washington, D.C. I can't really say where it is, it's on a need to know basis and you don't need to know where it is. You just need to know what it does. It's really just a huge array of state of the art mainframe computers containing a gargantuan database. This database is a continually updated record of the paper trail of classified documents. Not any of the classified material itself, mind you. But anytime a classified document changes physical location, or is moved to another controlling agency, or is shared with another agency, that transaction is recorded here. It's staffed by a small contingent of civilian contract employees, and if you saw it you'd think it was just another call center for The Franklin Mint or American Express. That is, until you tried to get through the door. Although I know its physical location, driven past it, in fact, I've never been inside. But I've called them many times.

I called them now.

"D.O.D. Classification Oversight," answered a bored voice. At least, it sounded like a real person instead of a computer.

I introduced myself, and then said, "I'm trying to locate some declassified files that I think are coming from Wright-Pat."

There was a short silence. Then, "Is that all the information you have?"

"Well, I think they were due to be declassified yesterday."

"Hmm. Well, we don't have reference documentation by base, or even by branch of service," he said. "As for a date, you had better run a three-day spread to make sure you catch the document and that gives you. . ." He paused and I could hear the rapid click of a computer keyboard in the background. "Approximately 6,478,923 pages. You do know about Exec Order 12598, don't you?"

I looked at my notepad again. "I think they are U.S. War Department documents, if that's any help."

"Most of this junk is War Department stuff," he answered. "You would be amazed at how much classified crap is laying around this country from World War II. Did you know that a toilet paper requisition order held a secret classification? It could be used by the enemy to determine troop strength of a particular outfit or garrison. I've got the entire Toilet Paper Requisitions here for Fort Wainwright from January 1942 to August 1974, if you want that."

"Thanks," I said. "I'll pass."

I was quickly despairing. Some mysteries, I supposed, were meant to remain mysteries.

ABOUT TO THROW IN THE TOWEL

I took one last look at my scribbled notes. The only thing I hadn't checked was a numerical sequence crowded into the lower left corner.

"Wait a minute," I said. "Do these numbers mean anything to you? C76497Q?"

"Sheesh!" he laughed. "Why didn't you tell me you had the invoice number?"

There was another rapid sequence of computer clicks.

"Here it is!" He said. "Came in yesterday! 15:32. Yep, it's a War Department reference, Procurement Board. Just put that reference number on your F.O.I. request. You'll want the file location."

I did, which he then gave me. It did indeed turn out to be at USAF Materiel Command at Wright-Pat. I went online, back to the AF Materiel Command home page, and clicked on "FOI Request." This is the document by which you request material under the Freedom of Information Act, or "FOI." I downloaded the PDF, and promptly filled it out, requesting access to all files listed on D.O.D. Record of Declassification Invoice #C76497Q. I dropped it in the mail and promptly forgot about it.

Three weeks later, a fat box was sitting on my porch when I came home. I took it inside and, after setting it on the kitchen table, split open the plastic tape with my best paring knife. The documents were, of course, new copies of the original files. Still, I couldn't help feeling as if the musty odor of 50 years in a dank storage vault wafted up out of that box. Laying on top of the bulging files was a copy of the declassification order: Form 12598. Section 2 of that order, "Military or Controlling Agency Name," answered all of my questions about who these files had belonged to.

According to the Declassification Form, the "Wardep" on my notepad was indeed the U.S. War Department. "ProcrBd" was the Procurement Board. This was the predecessor of the Materiel Commands of the five branches of the armed forces. "AeroDiv" was the Aeronautical Division of the Procurement Board; I was to later learn that it was responsible for evaluation of all aviation related purchases for all War Department Bureaus.

The "D.S.P." in my note indicated that the files belonged to the Department of Special Projects. I had never before heard of this agency. But, as I delved deeper and deeper into that box of obsolescent, arcane, and sometimes, just downright weird files, I would learn much about the Department of Special Projects in the months to come.

And what I would learn would change forever my understanding of the history of aeronautics. And it would change forever my understanding of the history of our nation. For these documents would ultimately reveal the bizarre way in which obscure and unknown agencies within our government work behind the scenes to accomplish their sometimes even more bizarre goals. Sometimes these agencies worked in harmonious concert; sometimes they worked at cross purposes, and sometimes they created more havoc for each other than the combined intelligence services of our most hated and feared enemies.

And I would learn that, sometimes, the cloaking of the functions of these agencies serves to protect the sovereignty of the United States, and the safety of its people. And sometimes, it serves to conceal a stupidity so colossal as to know no bounds.

J. Rutger Buck

Carmine de la Gracia, Colombia
August 26, 2001

WHO WAS ELMER C. WACKMALLIT?

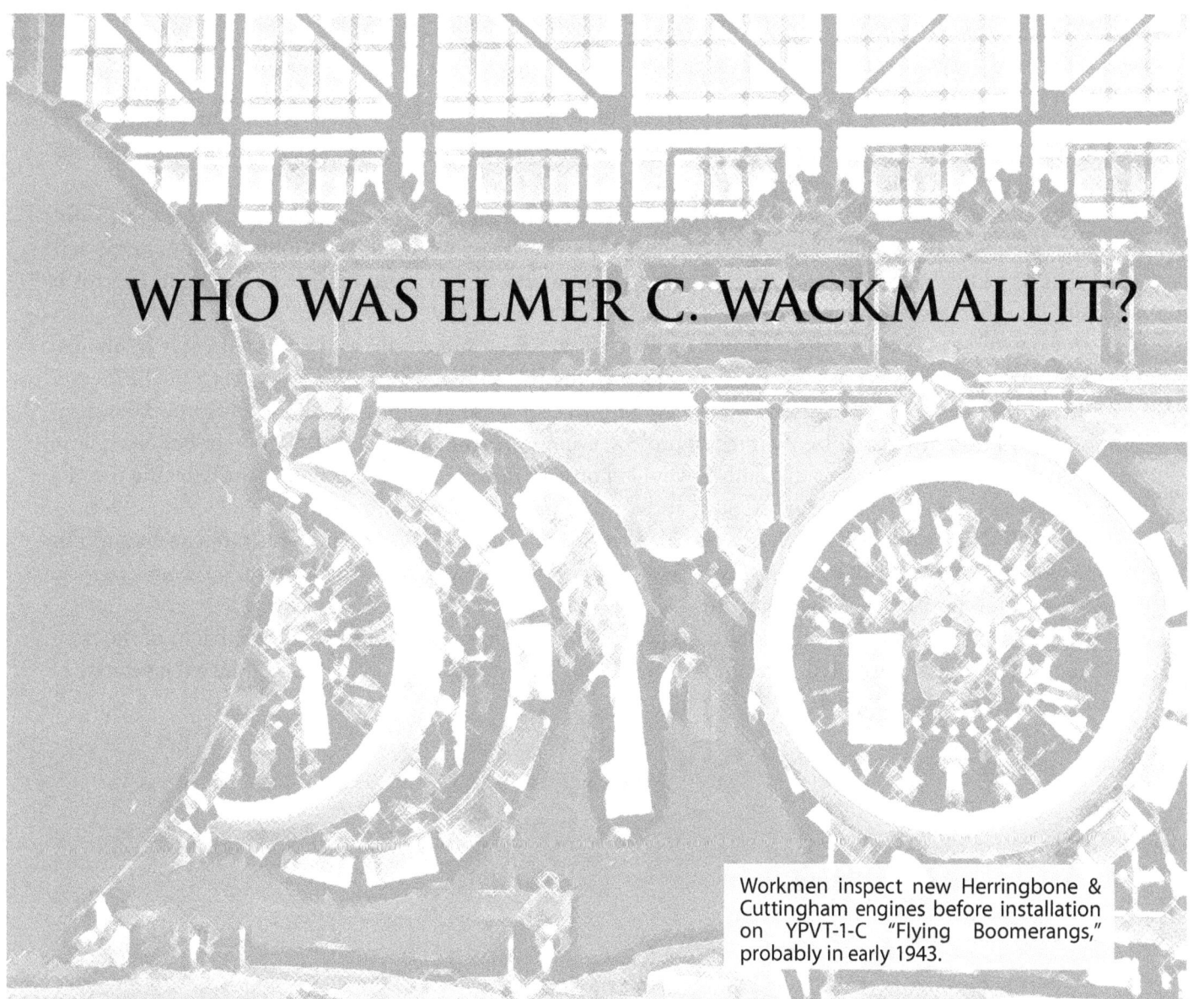

Workmen inspect new Herringbone & Cuttingham engines before installation on YPVT-1-C "Flying Boomerangs," probably in early 1943.

Arguably one of the most important documents contained in the files released by Declassification Order 12598, and thus brought at long last to light, is a lengthy manuscript entitled "The Flight of the Boomerang," an unpublished manuscript supposedly written by "Elmer C. Wackmallit." "The Flight of the Boomerang" is purportedly a work of fiction. However, it has been determined by Captain Buck, the author of "Assignment: Destiny!", to be a thinly disguised account of the actual development and deployment of the strange aircraft that was initially known as the Flungk Z-44 Mk I "Boomerang." The Z-44, according to Captain Buck's exhaustive research, is the same aircraft that later evolved into the YPVT-1C-EA "Indomitable Conflictor," an account that is indeed mirrored in the supposedly fictional account of Mr. Wackmallit. Both aircraft were referred to in common parlance as "The Flying Boomerang," which accounts for the title of Mr. Wackmallit's work of speculative fiction, "The Flight of theBoomerang."

Strangely, an exhaustive search of libraries, publishing houses, both major and minor, and the archives of all of the national publications over the last sixty years, has turned up not one single reference to an "Elmer C. Wackmallit." The United States Department of Social Security has no record of an "Elmer C. Wackmallit." Neither has the U.S. Internal Revenue Service, Army, Navy, Marine Corps, or Coast Guard, the Boy Scouts of America, the American Association of Retired Persons, or the International Order of Odd Fellows.

"Elmer C. Wackmallit" simply does not, and never did, exist. But, as we shall later see, the individual who, in all likelihood, was the actual author of "The Flight of the Boomerang," and employed the obvious code name of "Elmer C. Wackmallit," played a pivotal role in the procurement and subsequent development

1

of weapons systems for the Department of Special Projects. As such, he would have been privy to the inner workings of that super secret organization, and would have possessed detailed insider information about the projects that were developed within that organization.

These projects were all developed under a code of silence, a level of total secrecy, a cloak of mirrors, misinformation and outright deception so stringent, so all-encompassing, and so monumentally overpowering, that to even allude to the existence of such a program would have been regarded by those in control as nothing less than high treason.

In view of these facts, it is little wonder that the truth, fragile and elusive as it was, could only be alluded to in a work of obvious fiction, such as "The Flight of the Boomerang."

The author of such a piece surely would have known that those in a position of power within the organization would instantly recognize their secret project, even under the thinly veiled disguise of fiction. It is an even smaller wonder, then, that this author would choose to further obscure his true identity by the use of a fictitious pen name. Who is this mysterious author? You may well ask that question.

I submit that, at this point, at least, the true identity of "Elmer C. Wackmallit" is simply irrelevant. His name would be meaningless to the reader without important additional information; it is preferable to leave the revelation until later in this document.

Therefore, in the spirit of uncovering a deeper truth, the following excerpts from "The Flight of the Boomerang," this work of purported "fiction," are presented. The reader, as always, is left to judge the veracity of the information contained therein.

"THE FLIGHT OF THE BOOMERANG"

AN EXERCISE
IN DECEPTION

"Who, indeed, are we, to judge those who, when tried beyond endurance, revert in desperation to their last psychological line of defense against the cruel verities of fate, constructing in their tormented minds a reality within which they are, if not heroes, at the very least helpless and hopeless victims of a blind, uncaring, and vicious fortune over which they have absolutely no control?"

From: *Whackos, Crackos, Sickos and Psychos--A Psychological Study of Fruitcakes and the Nut-cases who Warped Them* by Dr. Waldo von Heinkerblonker, M.S., PhD, L.S.M.F.T. Out of print.

"You can always count on Dr. Waldo von Heinkerblonker to look on the bright side."

From: *They Might Have Been Giants: Misadventures, Blunders, & Colossal Failures in Aviation* by Halloway Bumpsteed, Jr.

"Oftentimes, when at long last the screen of deception is lifted, little remains but the framework of idiocy."

From: *The Veil of Stupidity: Code Level Ultra-Z Programs, Scams, Cover-Ups, and Outright Frauds,* by Dr. Eustis Bothomfieder

5

THE ESTEEMED SENATOR RECEIVES A COMMUNICATION OF THE UTMOST IMPORTANCE

(The following is an excerpt from *The Flight of the Boomerang*, by Elmer C. Wackmallit. Unpublished manuscript, Security Code Ultra {Level Z}, Office of Special Projects, Aeronautical Division, Procurement Board, U.S. War Department {Obsolete}. Released by Declassification Order 12598. Ref. No. C76497Q, Document # 3.)

Office of Senator Sylvester D. Quagmire
Senate Office Building, Washington, D.C.
August 21, 1942 9:47 a.m.

The Honorable Senator Sylvester D. Quagmire, R Iowa, settled his massive bulk into the rich brown leather of his new swivel office chair, all the while gazing fondly at the wall full of commemorative photographs, honorariums, and awards above his fireplace mantel. The good Senator appreciated a few minutes to reflect on his past glories each morning, the better to prepare himself for the rigors of the day ahead. And he appreciated his fireplace. He smiled to himself. Lots of Senate offices had a decent view, but in this day and age, precious few of them still had a working fireplace. The good Senator chuckled to himself, as he took a large red bandana from his coat pocket and wiped the sweat from his forehead. He wouldn't need a fireplace today. It wasn't yet nine o'clock, and it was already sweltering hot.

A sudden knock on the inner door disturbed the Senator's reverie. He sighed. The day had begun. He put his red bandana back in his coat pocket.

"What is it, Miss Castle?" he addressed the still closed door.

The door opened hesitantly, and Ruby Castle peered around the edge of the door.

"I'm sorry to disturb you, Senator, but these A-list correspondences came in the morning mail," she said hesitantly. "You said to let you know if we got any A-list correspondences, with it being an election year and all...."

"That's quite all right, Miss Castle," said the Senator amicably. "Let's see what you have."

The Senator grabbed the envelopes and quickly shuffled through them. From each envelope, he pulled forth a check, which he tucked into his center desk drawer. The Senator's "A-list" letters all came from constituents who had enclosed at least a two hundred dollar "campaign contribution." Each letter invariably contained a request, which his executive assistant Royce Jenkins would handle. Usually Jenkins would reply with a form letter designed to mollify the recipient. The form letter began: "As Senator of our great state I deeply share your concerns on the question of . . . . " Jenkins would then type in the appropriate response.

The Senator handed the letters to Miss Castle.

"Give these to Jenkins as usual, Miss Castle," Quagmire said. "Tell him to send the standard letter."

Miss Castle took the letters, but she didn't leave. She stood before the Senator's desk, a look of concern on her face. He noticed that her left hand was behind her back.

Senator Sylvester Bogmire, quite likely the real-life basis for Wackmallit's fictional Senator Quagmire.

"Miss Castle, is there something else?' he asked. He smiled his best election year smile.

"Tell me, what's on your mind?"

"It's . . . there's another letter, sir," Miss Castle brought her left hand from behind her back and handed the letter to Senator Quagmire. "It's not on the A-list, but . . . I really don't know what to make of it."

The Senator took the letter from Miss Castle. Then he sat back in his new swivel chair, and read the letter. As he read, he slowly leaned forward. When he was finished, Senator Quagmire held the letter out at arm's length between his thumb and forefinger. He stared at the letter as if it was diseased. Miss Castle couldn't tell whether the Senator was having trouble reading the letter, or somehow trying to distance himself from it.

"Senator?" Miss Castle prompted. The Senator shook his head, as if he was coming out of a trance. He handed the letter to Miss Castle.

"A crackpot, Miss Castle," he said, his smile icy now. "We get those letters once in a great while. Crackpot letters. From crackpots. Dispose of it. Please."

Miss Castle took the letter. Holding it with the pile of letters for Jenkins, she returned to her desk.

Miss Ruby Castle, real life secretary to Senator Bogmire, head of the Senate Defense Appropriations Committee, opens the Senator's files for the Senate Ethics Committee following Senator Bogmire's impeachment. The files confirmed the Senator's involvement in the "Boomerang Fiasco." (Photo courtesy of Dr. E.C. Whimpington)

Already it was unbearably stuffy in her little cubicle. Miss Castle laid the papers on her desk and turned on the big brass Bendix fan on top of the filing cabinet. At the same instant, the telephone rang. She rushed to pick it up. The telephone rang through to the Senator's office also, and she didn't want the ringing to disturb him. The poor dear, he worked so hard.

"This is the office of Senator Quagmire. How may I help you?" she said into the telephone.

The big brass Bendix fan picked up speed. It began to rotate around the room. It blew across the desk. All of the letters flew into the floor. Miss Castle, professional that she was, maintained her demeanor until she finished the call. Only then did she pick up the letters from the floor. She placed them all in a manila envelope. On the front of the envelope she wrote, "T. R. Jenkins, Room 1412." She dropped the envelope into the OUT basket.

Shortly after lunch, an office boy came by. He dropped the afternoon mail in the IN basket. He picked up the mail from the OUT basket, including the manila folder for Jenkins.

THE CONSTERNATION OF
T. ROYCE JENKINS

Office of T. Royce Jenkins
Aide to the Honorable Senator Quagmire
Room 1412
Senate Office Building, Washington, D.C.
August 21, 1942. 3:18 p.m.

T. Royce Jenkins sat at his desk, his coat hung over the back of his chair. His little Crosley fan rotated back and forth, rearranging the sweat patterns on his forehead. He read the letter a second time. Then, he read the letter a third time.

"Dear Senator Quagmire:

"In this time of national crisis I feel it is my duty to contact you on a matter of the utmost importance to our future as a nation. Although I have been sworn to secrecy in this matter, I feel a higher duty impels me to write you.

"You may not be aware of this fact, but it has been my habit for some years now, to engage in the consumption of tin foil. This started quite by accident in 1928, when I ingested a pastry from an auto-mat in which the tinfoil wrapper had been squashed into the cake. I found the tingling chemical taste of the tinfoil quite pleasurable. I naturally increased my consumption, until one evening while walking home from the auto-mat, my mouth full of tinfoil, I was struck by lightning.

T. Royce Jenkins, acerbic aide to the Honorable Senator Sylvester Bogmire. A faithful contribution collector for the esteemed Senator, Jenkins was nevertheless perplexed by Smith's bizarre request. (Whitney Speale Collection)

"Since that time, as it might amaze you to know, I have been in almost constant contact with the Zenturions, from the planet Zenturia, who coincidentally also chew tinfoil!

"As one of the planet Earth's most advanced beings, it is only natural that the Zenturions would contact me. They beam their messages to me across thousands of light years instantaneously through a device they call the "Radioteleportatron."

"Of course, I have informed the Zenturions of our current difficulties with the Axis powers. I hope you don't mind. They have graciously offered their assistance. For the last six months they have been

"Xontar the Magnificent" Smith in full ceremonial regalia. He is standing in a washtub full of tinfoil, which he believed would increase his receptive powers. (Photograph courtesy of Dr. Waldo von Heinkerblonker, M.S., PhD, L.S.M.F.T.)

beaming me detailed instructions for the construction of a Trans-Dimensional Space Cruiser, equipped with the very latest Zenturion Death-Ray.

"This machine will guarantee our absolute aerial superiority in every theater of war on Earth, not to mention the entire Solar System! I am currently building the prototype in my garage. It will require very little in the way of strategic war materials, being constructed almost entirely out of cardboard boxes and used tractor parts!

"I will be ready for testing very shortly. Could you get in touch with the proper military authorities and have them present for the test flight? I anxiously await your reply.

"Sincerely:

Fisher P. "Xontar the Magnificent" Smith

1654 34th Avenue

Council Bluffs, Iowa

"P.S. Please do not come at night. Due to the sensitive nature of this enterprise, I must shoot first and question later. I do hope you understand."

Jenkins could only shake his head in wonder. The old man must have gotten a doozy of a "contribution" from this crackpot for him to merit "special handling." No problem. Jenkins rolled a clean sheet of paper into his typewriter and started off with form letter number one:

"Dear Mr. Fisher P. "Xontar the Magnificent" Smith:

"Thank you for your recent letter. Your concern is of the utmost importance to me. As Senator of our great state I deeply share your views on the question of Trans-Dimensional Space Cruisers, equipped with the very latest in Zenturion Death-Rays. I understand your interest, for I, too, have often contacted aliens while chewing on tinfoil. . . ."

Jenkins pulled the paper from the typewriter, wadded it up and tossed it in the trash. He looked at the trash can, had second thoughts, and retrieved it .He dropped the wadded paper into the ashtray on his desk.

Then he pulled out his Zippo and lit the paper. It would be a disaster to have even this partial reply seen by anyone.

This would never do. Stronger measures were in order. Jenkins picked up the telephone.

"Alice, get me Hackett in the War Department," Jenkins said. He impatiently tapped his fingers on the desktop, while waiting for the connection to be made.

The intercom buzzed on Jenkins' desk.

"Colonel Hackett's on the line, Mr. Jenkins," Alice cooed into the receiver and clicked off.

"Hackett!" said Jenkins brightly into the telephone. "Yes! Fine! How are the kids? Good, good! Say, remember that deal last March, when I used Quagmire's pull to get that black budget appropriation for your Special Products thingamajig through Congress?

"Yes? Well, whatever you call it, Department of Special Projects, whatever. Yes, that's it, the one where you said 'If there's ever anything my office can do for you?' Yeah, yeah, anyway, that's the outfit where they're looking for new aircraft, right? Well, give me their address. Boy, oh, boy, have I got a lead for those guys . . ."

2. THE BUCK IS PASSED

Senate Office Building, Washington, D.C. Repository of the office of the Honorable Senator Sylvester D. Bogmire, upon whom the fictional "Senator Quagmire" of Wackmallit's story is based.

(Excerpt from *The Flight of the Boomerang*, by Elmer C. Wackmallit, continued.)

Department of Special Projects
Office of Procurement; U.S. War Department
Washington, D.C.
August 22, 1942 2:36 p.m.

The August air hung like a hot and humid veil over the city. Scarcely a breeze moved. These were the days before air conditioning, and in the vast halls of bureaucracy, every window that wasn't painted shut was opened wide.

The mail clerk propped his bicycle against the nondescript brownstone building. Pulling an already sopping handkerchief from his back pocket, he mopped it listlessly across his forehead. He grabbed his mailbag from the front basket and hung it over his shoulder. Then he walked through the massive wooden front doors and into the hallway, out of the scorching mid-afternoon sun.

Once inside, what seemed like an ordinary apartment building from the outside was magically transformed. The well-lit hallway was bustling with activity. Uniformed non-coms crossed paths with smartly dressed civilian secretaries, all carrying piles of file folders.

The messenger, though, had seen it all before. Besides, he was dying of thirst. He went to a water fountain down the hall and turned the chrome handle. A feeble jet of tepid water squirted from the spout. He let the water run for a bit. But it didn't cool off, it got hotter. Still, it was wet, and it gave him enough renewed vitality to climb the three flights of stairs to the office of Col. J. Bristle Davenport, Office of Special Projects, U.S. War Department.

As the messenger boy stepped onto the third floor landing, he heard a raucous din emanating from behind the closed door of the Department of Special Projects. Even in this oppressive heat, the D.S.P. door was closed. He walked down the

11

hallway, cautiously, toward the closed door.

"All you've got? What do you mean, this is all you've got?" a gravel voice boomed angrily, rattling the glass in the door. "Isn't there a single aeronautical engineer in this entire country who can come up with an original idea?"

The messenger knocked hesitantly. Cigar smoke rolled out in waves from under the door.

"You've combed through the design department of every major airplane manufacturer in the United States," the din continued, "and the best you can come up with is a proposal for a twin-engined P-40 and a retrofit turbo-charger for the Brewster Buffalo?"

The messenger brought up his hand to knock again. Instead, he almost rapped a quickly exiting Army Air Corp Major in the face with his knuckles. The messenger walked hesitantly into the office. The Major scurried down the hallway.

Ironically, despite his "tough guy" demeanor, as a test pilot, Davenport was prone to bouts of airsickness. He is shown here after a barnstorming flight, recovering from the physiological effects of stunt flying. (Image courtesy of the private collection of Dr. Manfredd von Goetzzenberger of Das FleugelWerkes)

In the years before the war, Davenport's work as a civilian contractor with the Army Air Corp Specifications and Acceptance Board at Wright Field, Ohio led him to evaluate the Boeing XP-936 which resulted in the Boeing P-26 Pursuit aircraft, (Pictured inset left) and the Curtiss XP-36 (Pictured above and inset right). The P-36 was a precursor to the vaunted P-40 of "Flying TIger" fame. (Photographs courtesy of the U.S. Air Force Museum)

Col. J. Bristle Davenport leaned forward, his bald dome gleaming, his fists planted on the desk, and his cigar clenched between his yellowed teeth in a death-grip snarl.

"And don't come back until you've got something worthwhile!" He puffed. His beady eyes malevolently swiveled to the messenger boy, seeing him for the first time. "And who might you be, sonny?"

"Uhh," the messenger stammered. "Special messenger, sir. I mean Colonel, I mean, uhh, Colonel Sir. . . . Sir."

The messenger boy tossed a manila envelope onto the Colonel's desk. "From Senator Quagmire's office, Sir."

The Colonel looked down at the envelope sliding across his desk. Quagmire . . . Quagmire . . . Now where had he heard that name before? Oh, yes, Senate Defense Appropriations Committee. He would have to pay special attention to this one.

Davenport ripped open the letter and quickly scanned the contents. The messenger boy stood, ill at ease, waiting.

After some minutes, Colonel Davenport slowly raised his head. An evil smile crossed his weather beaten face. He looked at the messenger, then gestured with his head down the hall, in the direction of the scurrying Major.

"Is that Major still out there?" he asked the messenger.

The messenger stepped backward to the doorway. He saw the Major at the far end of the hallway, nervously tapping a Chesterfield out of a pack and trying to light it. His hands were shaking so much that he couldn't strike a match.

The messenger boy shook his head dully at the Colonel, indicating the affirmative.

"Speak up, boy!" yelled the Colonel, puffing out a tremendous cloud of cigar smoke "Don't you know there's a war on?"

"Yessir, Colonel Sir!" stammered the messenger. "I mean, he's out there, Sir!"

"Claproot!" The Colonel bellowed, rattling the door glass once again. "Get back in here! I've got just the project for you!"

COLONEL J. BRISTLE DAVENPORT: IRON BACKBONE OF THE D.S.P.

Col. J. Bristle Davenport was old school Army Air Corp all the way. In 1917, weary of slugging it out in the trenches, he had talked his way into the Army Expedition Force as an aviation lineman, gradually working his way up to maintenance officer. He cadged flight lessons where he could, finally making it into the cockpit. He ended the war with six huns to his credit, including one Fokker DR-1 from J.G. 1, Richthofen's famous Flying Circus.

In the lean years between the wars, he left the service and worked many aviation odd jobs, including a not so well-paying stint at barnstorming. He eventually wound up as a line mechanic for T.A.T. airlines, wrenching on Ford Trimotors and Fokker F-2Bs all night, while attending college in the day. He received his degree in aeronautical engineering and applied for employment in his field. Six weeks later he was hired as a civilian contractor and attached to the Army Air Corp Specifications and Acceptance Board at Wright Field, Ohio. His first assignment was evaluating a new Boeing model designated the XP-936, beginning in March, 1932. He remained at the facility for almost eight years, performing static trials, fitness for service trials, and squadron trials on prospective aircraft for service with the U.S. Army Air Corp.

When war was declared, he reentered the service with the rank of Captain, and was posted to the U.S. War Department Procurement Board in Washington, D.C. When the order came down to establish the Office of Special Projects, Davenport, because of his extensive background in aircraft evaluation, was given the brevet rank of Major and chosen to head the group.

THE BUCK IS PASSED, AGAIN

(Excerpt from *The Flight of the Boomerang*, by Elmer C. Wackmallit, continued.)

Office of Major Wendell Claproot, Room 628
Department of Special Projects
Office of Procurement; U.S. War Department
Washington, D.C.
August 22, 1942 4:17 p.m.

"Lacksdale!" Major Claproot stared down at the envelope on his desk as he bellowed into the anteroom outside of his office. "Get in here! I've got just the project for you!"

THE BUCK STOPS

(Excerpt from *The Flight of the Boomerang*, by Elmer C. Wackmallit, continued.)

Office of Captain Rodney Lacksdale, Room 306
Department of Special Projects
Office of Procurement; U.S. War Department
Washington, D.C.
August 23, 1942 7:16 a. m.

"Blakeley!" Captain Lacksdale held the manila folder before him as he bellowed into a burgundy bakelite intercom on his olive drab metal desk. "Get up here! I've got just the project for you!"

Five minutes later, Blakeley stood before Captain Lacksdale's desk. Blakeley, in his early thirties, was an unlikely warrior. Slight, balding, with round horn rimmed glasses, Blakeley had resigned from his position as head (and only) aeronautical engineer at Empire Aircraft Company the day after Pearl Harbor. That same afternoon, he volunteered for service.

For many, this would have been a patriotic act. For Blakeley, it might have been patriotic, but it was also an astute career move.

Ironically, Empire Aircraft Company would go down in history as the only aircraft Company to go bankrupt during the war years. This was a prodigious feat when every aircraft plant was operating at over-full capacity. Even refrigerator manufacturers were building dive bombers. The Empire Aircraft Company accomplished this unique feat primarily by the stubborn insistence of the Company President that all aircraft be built out of stainless steel because it was "nice and shiny."

Blakeley, as head engineer, had tried in vain to point out that, while being nice and shiny, stainless steel nonetheless weighed half again as much as the new aluminum alloys coming into use.

It was to no avail. Empire continued to produce gorgeous airplanes that looked good on the flight line, but could barely get off the ground.

Above, then Lieutenant Rodney Lacksdale is seen displaying his newly issued Sam Browne belt in 1927. Although his hickory pointer is absent in this photograph, the air of self-satisfied officiousness which would serve him so well in later years is already evident. (Photograph courtesy of the Whitley Speale Collection of the Bone Lake Research Museum)

So far, Blakeley had been very content with his newest career move. Only now, he was staring incredulously at his orders. "You can't be serious, sir."

Lacksdale continued writing. Without looking up, he said, "You have your orders, Lieutenant."

"But, Captain," Blakeley sputtered. "These orders say to drive all the way to Council Bluffs, Iowa? Can't you get me a transport requisition?"

"You know better than that, Lieutenant!" Lacksdale said. "With the big push in North Africa, all the air transport traffic this side of the Mississippi is headed eastbound. Besides, how many flights do you think the Transport Corps has to Council Bluffs, Iowa? You'd be lucky to cadge a flight in weeks, and we can't wait that long. There's a war on, you know!"

Blakeley looked down at his orders. He shuffled through the papers, slowly at first, then almost frantically.

"Anyway," said Lacksdale. "you gotta go. The old man says that this guy is a personal friend of Quagmire's."

Blakeley looked up from the orders, staring at Lacksdale in disbelief.

"Did you read this, Captain?" he asked, incredulously. "I'm driving four days out to the middle of the country to look at a tractor powered cardboard and tinfoil spaceship?"

"Be nice to the man, Julian," said Lacksdale. "Even if he is nuttier than a fruitcake. Just go on out there and look at his spaceship. Then just tell him we'll be in touch. Remember, Quagmire is head of Senate Appropriations. I don't need to tell you he's the guy who signs all our paychecks."

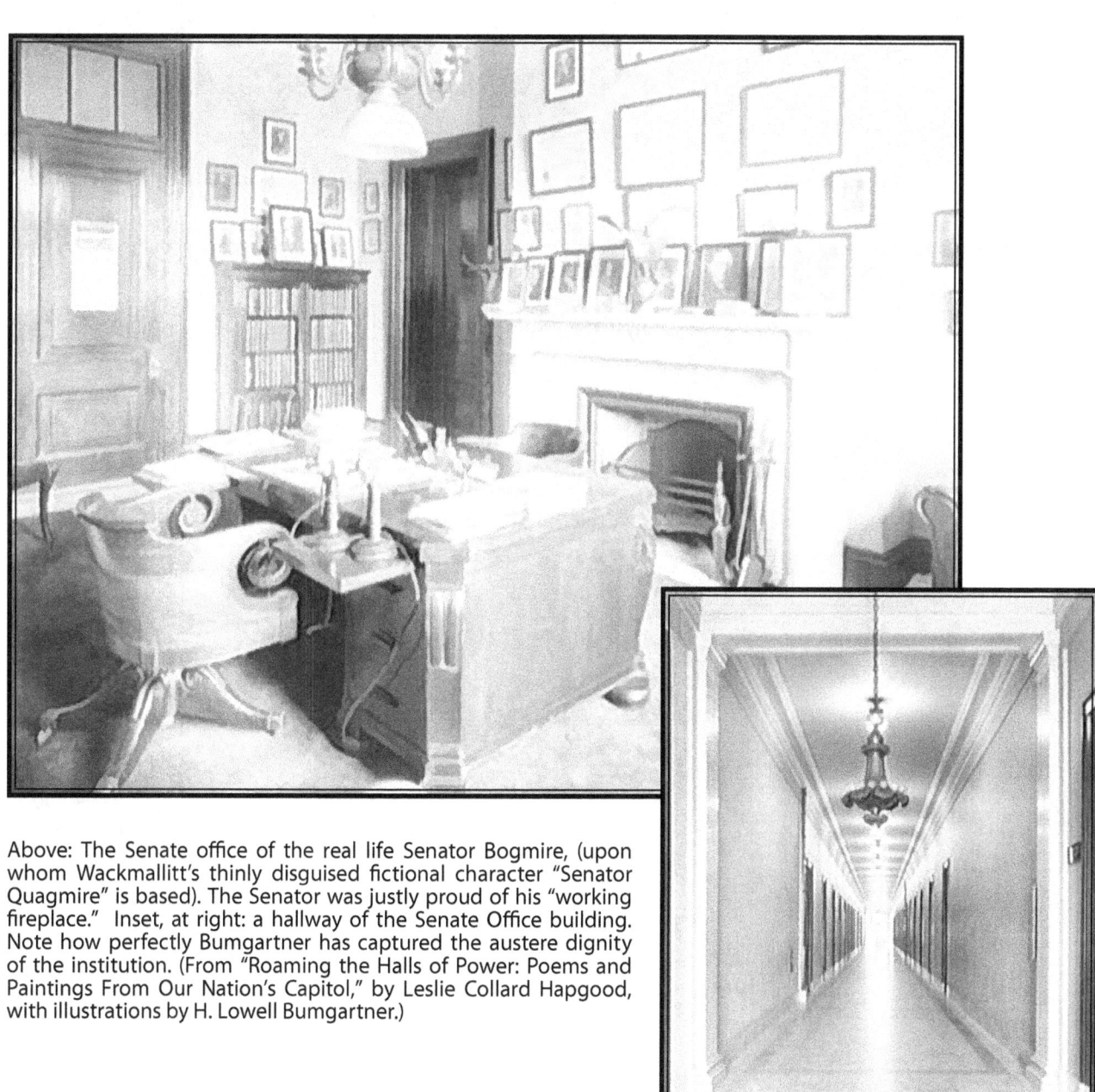

Above: The Senate office of the real life Senator Bogmire, (upon whom Wackmallitt's thinly disguised fictional character "Senator Quagmire" is based). The Senator was justly proud of his "working fireplace." Inset, at right: a hallway of the Senate Office building. Note how perfectly Bumgartner has captured the austere dignity of the institution. (From "Roaming the Halls of Power: Poems and Paintings From Our Nation's Capitol," by Leslie Collard Hapgood, with illustrations by H. Lowell Bumgartner.)

The questionably famous, but undoubtedly notorious, "House of Xontar," on the outskirts of Council Bluffs, Iowa. Besides being Lieutenant Julian Blakeley's intended destination and a handy locale for contacting your dead relatives, the "House of Xontar" was the repository of "Xontar the Magnificent" Smith's collection of scrap iron and used plumbing fixtures. (Photograph courtesy of Halloway Bumpsteed, Jr.)

3. A GREAT JOURNEY BEGINS WITH A SINGLE STEP

The U.S. War Department Building, Washington, D.C. The nexus of power over all things military, the home of the Department of Special Projects, and the starting point for Lieutenant Blakeley's momentous journey. (Photograph courtesy of the Harris & Ewing Collection of the Library of Congress.)

(Excerpt from *The Flight of the Boomerang*, by Elmer C. Wackmallit, continued.)

Office of Lieutenant Julian Blakeley, Room B14
Department of Special Projects
Office of Procurement; U.S. War Department
Washington, D.C.
August 23, 1942 8:14 a. m.

Lieutenant Julian Blakeley, lifted the pile of documents from his desk, and proceeded to the motor pool. He gave the Sergeant at the motor pool desk his Form 4145 Dispatch Record, and the Sergeant handed him a set of Ford keys.

"Lot 3, Space 44, Lieutenant," the Sergeant said, saluting smartly. It was still Washington, after all. "Have a nice trip, sir."

Blakeley, keys in one hand and files in the other, attempted a salute. Instead, he dropped the keys to the Ford. Bending over to pick up the keys, he dumped the document pile in the floor. Then he picked up all of the files. Finally, laying his file folder and the keys both on the desk, Blakeley smartly returned the Sergeant's salute. He then gathered his folder, orders, and car keys, and walked out into the motor pool. At the desk, the Sergeant shook his head sadly.

Blakeley found the car in the motor pool parking lot, a 1941 olive-drab Ford four-door sedan with a white star painted on the driver's door.

After carefully stowing his orders and documents, Lieutenant Julian Blakeley pulled out of the motor pool, turned left, and drove down Twelfth Street, where he turned right onto the wide expanse of Massachusetts Avenue. The traffic was still heavy here in the city, even with war rationing in effect. In fact, Blakeley even encountered a slight slow down in DuPont Circle, that in these wartime days would count as a traffic jam.

However, as Massachusetts Avenue became Route 240, the traffic thinned somewhat. By the time Blakeley turned north on Route 355 toward Rockville, with the exception of an occasional passing freight truck, he had the entire road to himself.

Although gasoline rationing was in effect throughout the war, Washington, D.C. traffic was at times moderately heavy. (Office of War Information photo.)

LT. BLAKELEY ENCOUNTERS
A SMALL DETOUR

By mid-morning, he was crossing the Pennsylvania state line. Still, there was almost no traffic. Then, just outside of Uniontown on Route 40, he heard the highway sound that no one wants to hear. When he peered in the rear view mirror, he saw the blinking red lights of a Pennsylvania State Motorcycle Cop.

Blakeley pulled the Army Ford over, and the cop skidded in behind him. Sliding his goggles up over the bill of his peaked cap, the Motorcycle cop dismounted and leaned the big Harley on its kick-stand, simultaneously opening his ticket book. He swaggered up to the Ford's driver window.

At first he tried to keep to the wartime posted speed limit of 35 miles per hour. Outside of Hagerstown, though, he figured out that at this pace, he wouldn't make it to Council Bluffs before the war was over. So, he sped up to forty-five, then fifty, then sixty. The flat-head V-8 was screaming. It had finally occurred to Lieutenant Blakeley that, after all, his mission was what all the gas-saving and tire-saving was about, wasn't it? After that, he made good time, cutting through the rest of Maryland at a healthy clip.

"Alright, General," the trooper pulled a pen out from his shirt pocket. "Where's the big emergency today?"

"Officer," Lt. Blakeley smiled as pleasantly as he could, "I know you think that you are doing your

Before joining the Pennsylvania State Troopers, Trooper Corbett was a motorcycle patrolman for the City of Pittsburgh Municipal Police Department. He is shown here on his service Harley Davidson in 1932. (Photograph courtesy of the Harris & Ewing Collection of the Library of Congress.)

duty, but I'm on official business for the United States Army Air Corps."

"Oooohhh, official business?" The trooper tilted back his hat with his pen. He smiled, a huge toothy grin. Blakeley expected to see a shining glint of light reflect off of his incisor at any moment. "Top secret, I'll bet!"

"As a matter of fact, yes, officer, it is," said Blakeley cordially. "So if you'll excuse me, I'll be on my way."

"Oooohhh, in a big hurry are we?" The trooper's grin got even bigger. "And where is this top secret mission that you're in such a hurry to get to? I mean, that is, if you can tell it to me, being just a lowly civil servant and all."

"I don't suppose it would hurt," said Blakeley affably. "It's in Council Bluffs, Iowa. Now, if you'll excuse me-"

"Council Bluffs, Iowa?" said the trooper, his eyebrows raising ominously. "And what have we got there? Japs? Or Krauts? Or maybe both?"

"Neither, really," Blakeley laughed uncomfortably. "You see--Well, I really can't tell you, because it's--secret, see?"

"Oh, I see!" said the cop pleasantly, nodding his

THE SHORT BUT HEADY LIFE OF THE "AERIAL FORTRESS" CONCEPT

Throughout the twenties and thirties the concepts that would define the U.S. Army Air Corps' role in the defense of the nation were rapidly being developed. Advocates of strategic bombing strongly believed that massive bombers armed with defensive weapons were the wave of the future. General Billy Mitchell's highly publicized sinking of battleships in July, 1921 by bombs dropped from aircraft demonstrated that air power was a stand-alone strategic force. No longer could airplanes be considered as simply a supporting element to traditional land and sea forces.

In response, the U.S. Army Air Corps belatedly called for the development of long range bombers capable of defending themselves from attack by enemy fighter aircraft. The theory was simple enough: put enough defensive machine guns on the

The Douglas XB-19 dwarfs men and equipment in the foreground. Note the Douglas DC-3's relative size in the background. (Photograph courtesy of the U.S. Air Force Museum)

long range bombers so that they could defend themselves.

Additionally, the strategic thinking of the time envisioned that these aerial behemoths would be based in the U.S. Therefore they should have tremendous range in order to reach enemy targets in either Europe or the Pacific, and be able to reach targets in Hawaii, Alaska, and the Panama Canal Zone.

So in 1933, the U.S. Material Division at Wright Field issued a specification for a strategic bomber capable of carrying a 2500 pound bomb load to a target 5,000 miles away at a speed of at least 200 MPH.

In response, three companies developed concept aircraft for the Army Air Corps' XBLR (Experimental Bomber, Long Range) program. The entry of the Glenn L. Martin Company, the XB-16 was soon deemed too expensive and cancelled. Two other entrants, the Boeing XB-15, and the Douglas XB-19 were each subsequently built as a single model.

The Douglas XB-19 flew on June 27, 1941. It had a range of 7,300 miles with a 6,000 pound bomb load, but its cruising speed was a pokey 135 MPH.

The Boeing model, the XB-15, flew on October 15, 1937. While it had a higher cruising speed, 171 MPH, than the Douglas entry, its range was only 3,400 miles with a 2,500 pound bomb load.

Neither model proved adequate for production. The lag time between the contract specifications and the actual flight of the two aircraft virtually insured

The Boeing XB-15 in flight. Note the pre-1942 wing roundels and tail stripe. (Photograph courtesy of the U.S. Air Force Museum)

head eagerly in understanding. "I get it! You can't tell me because it's--secret, right?"

"Yeah, that's it!" Blakeley beamed. He pushed in the clutch and pulled the gear lever down into first. "Well, I gotta go-"

"Just a minute!" The trooper put his hand on the window sill of the Ford. His face clouded ominously. "If there are Japs or Krauts in Council Bluffs, pretty soon they'd be in Peoria, then maybe they'd invade Indianapolis. Next thing you know, there they are in Columbus, which is just down the road from Wheeling, which is practically next door to Pittsburgh, which is just a stone's throw from Uniontown, which happens to be where my detachment is. And you think I don't need to know about that?"

"Look, for Criminy's sake!" Blakeley sputtered, exasperated. "There aren't any Japs and there aren't any Krauts! But you are standing in the way of official United States Army Air Corps business!"

"No Japs and no Krauts, eh?" The trooper's eyebrows raised once again. "I guess Tojo and Hitler are just figminks of my magikanation."

"What?" Blakeley stared at the trooper. "Do you mean figments of your imagination?"

"Yeah!" the trooper scowled. "That's what I said, ain't it? Figminks of my magikanation!"

"I've had about just about enough of this," said Blakeley, growing irritated at this unwanted interruption.

The trooper pulled open the car door, simultaneously drawing his Smith & Wesson K-38 police revolver. "Get out of that car! Lemme see your gas ration card!"

"I don't have a ration card!" Blakeley said, stepping out of the car and raising his hands above his head. "We're the United States Army! We're who you're rationing for."

"Ohh, you wise guys crack me up! Suppose you tell me where you got them brand spanking new tires? Just you look at my bike!" The motorcycle cop pointed back at his motorcycle. "Even I have to ride on old baldies back there. My rear tire is as slick as a newborn baby's butt! And me, a public servant!"

"Look," said Blakeley. "I just picked this car up out of the motor pool-"

"How do I know you're a soldier, anyway? Maybe you could even be one of them Germans!"

"I thought maybe the uniform would give you a hint." asked Blakeley.

"Uniform?" exclaimed the trooper incredulously.

"Who are you trying to kid? Cripes, even my

that both the XB-15 and the XB-19 were obsolete well before their first flights. When hostilities broke out, both aircraft were converted to cargo planes.

In 1942, Alexander P. de Seversky's influential book, "Victory Through Airpower," was published. In this book, and a subsequent film of the same title produced by Disney Studios in 1943, Major de Seversky, a pilot and aircraft designer, strongly advocated for the type of strategic air power as envisioned in the thirties, with heavily armed aerial fortresses penetrating deeply into enemy territory to bomb strategic targets.

Unfortunately, events in Europe would soon prove that strategic bombers over enemy territory were highly vulnerable to enemy fighter aircraft, regardless of how heavily they were armed for self defense.

The Air Corps top brass who had been fixated on strategic bombers were now clamoring for fighters capable of defending those very bombers from the enemy. They naturally looked to the Department of Special Projects, leading directly to Lieutenant Blakeley's ill-fated road trip.

A ramp shot of the Boeing XB-15 gives a good indication of its proportions. A problem with these gigantic bombers which seems obvious in retrospect, is that, given their extreme ranges and slow airspeeds, two and sometimes three crews were required for each aircraft. The loss of a bomber would therefore mean the loss of multiple crews. Note the spectators gathered in the shade of the huge wing. (Photograph courtesy of the U.S. Air Force Museum)

little nephew Nemo could walk into the Quartermaster store on State Street in Pittsburgh and get that whole outfit, from your shiny brown oxfords all the way up to your little soda jerk hat! In fact, that's just what he did! Says the dames go wild over a guy in a uniform."

Blakeley reached up with his right hand. The trooper crouched in a shooting stance and cocked his revolver. "Watch it with the fast moves, buddy!"

Blakeley slowed his movement, and carefully pulled out his dog-tags from beneath his shirt. "Here, you can check my dog-tags."

"They got a machine that makes them up at the same store. Twenty-nine cents apiece. Little Nemo's got a pair of them. They're only forty-nine cents if you buy two."

"Officer, you've got to believe me! National Security is at stake! Isn't there anything I can do to convince you?" asked Blakeley.

"Oh, it isn't me you've got to convince," said the trooper. "Get back into your car and follow me into town. I'll show you where to pull over. Then you can tell your story to the county magistrate."

Blakeley got back in the car as the trooper walked back to his motorcycle.

"And remember," yelled the trooper over his shoulder. "No funny stuff, see? I'm right behind you!"

KEEP 'EM FLYING!

U.S. ARMY AIR CORPS

4. BLAKELEY'S DAY IN COURT

Every seat in the courtroom was filled for Lt. Blakeley's trial.

(Excerpt from *The Flight of the Boomerang*, by Elmer C. Wackmallit, continued.)

Fayette County Courthouse
1212 North Court Street
Uniontown, Pennsylvania
August 23, 1942 2:46 p. m.

Magistrate Warren T. Cockthorn pulled contentedly on his plaid suspenders and leaned back in his ancient wooden office chair. The chair creaked audibly in the high ceilinged courtroom.

The Magistrate sat grandly, in a manner appropriate to his position, behind an ancient rickety wooden judicial bench. The bench itself sat before the jury box in the front of the courtroom, properly elevating the judicial bench slightly above the courtroom galley, as befitted the august station of the Magistrate.

"Ninety days labor on the county farm, five dollars fine and eleven dollars in costs!" intoned Magistrate Cockthorn, as he banged his gavel without too much enthusiasm on the creaky wooden judicial bench.

The constable, standing before him and holding the elbow of a scruffy lout, in a red checkered shirt and sporting a couple of days beard growth and an obvious hangover, leaned forward by the side of the desk and whispered something audible only to the judge.

"Henry," the magistrate almost whined the name, pulling it out to three syllables. "We've been over this a thousand times before. I told you that you don't get a cut of the labor, that's a county farm deal! And the costs belong to the court, which you may have noticed, happens to be me! You'll get your share

23

Magistrate Warren T. Cockthorn, "The Law West of the Youghiogheny."

file folder. The magistrate idly drummed his fleshy fingers on the table. A couple of old men, courthouse loafers, sauntered into the courtroom, where they took seats in the back. Blakeley gritted his teeth. The trooper glanced around the courtroom. No bailiff was in sight.

Blakeley looked once more at the Magistrate, who was staring idly at the ceiling, where a huge old fan silently turned. The trooper half turned to Blakeley, and almost sheepishly looked at his prisoner. Finally the trooper stood, coughed into his fist, and addressed the Magistrate.

"Umm, your honor," he said. "When we came in, I believe we passed Bailiff O'Riley coming out of the front door of the courthouse."

The Magistrate sighed. "I suppose he's over at Milligan's Bar again. Well, far be it from me to let work stand between an Irishman and his bottle. What gross dereliction of the legal system of this great country do you have for me today, Trooper?"

"Your honor," said the trooper, "this culprit passed me on Route 40 going at least eighty miles an hour. It was all my Harley could do to catch up with him. I was headed eastbound just crossing the Youghiogheny Bridge, when he passed me going

of the fine like you always do. And besides, the open court is not a fit forum to discuss the economics of the County judicial system." He banged the gavel again, harder this time. "Now, remand the prisoner to the city jail for transportation to the county farm like you're supposed to do!"

The constable guided the prisoner out the side door. The magistrate turned his head to the right, expertly spitting a stream of tobacco juice into a nearly overflowing tomato juice can, which sat on the floor beside him, beneath the desk.

The magistrate, being a better litigator than housekeeper, had amassed quite a collection of tomato can spittoons beneath his desk. He wasn't sure of the exact number, but each time he filled one up, he pushed it back under the desk with the toe of his Frye boot. To the best of his recollection, there was space under the desk for seven or eight tomato cans. He would have to remember to get some county labor up here and do some judicial housecleaning. Cockthorn ran the fleshy fingers of his right hand through his grey hair, adjusted his glasses, leaned forward, and said, "Bailiff! Next case."

Blakeley sat morosely on a wooden bench beside the motorcycle trooper, clutching his orders and

The county officer's badge worn by Warren T. Cockthorn during his illustrious and lengthy tenure as Magistrate for Fayette County. (Photograph courtesy of Warren T. Cockthorn III and the Fayette County Historical Society)

westbound like a bat out of Hades, excuse my expression."

Blakeley jumped up from the bench. "That's an outright lie!" he shouted. "I wasn't going a mile over sixty. Plus, I resent being called a culprit!"

The Magistrate rapped the gavel on the bench three times in rapid succession. "Sit down, soldier boy!" he said. "You'll have your say in court."

The two grizzly old-timers in the back of the courtroom looked at each other, smiling in anticipation. One leaned over and whispered to the other. The second old-timer got up and hurried out of the courtroom door.

"Anyway," stated the trooper, "by the time I turned around, I could just see him going over the rise. I probably never would have caught him if he

DEAD ENDS AND FRUITFUL PURSUITS: THE STATE OF PURSUIT AIRCRAFT DEVELOPMENT IN PRE-WAR U.S.A.

In the years preceding the United States' entry into World War II, there was an intense national debate concerning the U.S. role in what many perceived to be a European War. While President Franklin D. Roosevelt favored aiding the British in the defense of their homeland, there was a significant faction of the American population who either favored isolation from involvement in European affairs, or who actually favored the Nazi regime of Adolf Hitler.

The Curtiss XP-46 was conceived in 1939 as a replacement model for the Curtiss P-40, which was at that time a front line Air Corps fighter. When flight trials demonstrated that the XP-46 was in fact inferior in performance to the P-40, the new fighter was scrapped. (Photographs courtesy of the U.S. Air Force Museum)

Foremost among these isolationists was no less a personage than Charles H. Lindbergh, perhaps the only man in the world with as much prestige, and some said, more prestige, than the President himself. After visiting the German Luftwaffe in 1938 at the request of the Army Air Corps, Lindbergh became convinced of the superiority in all respects of the German aircraft, production methods, and military organization.

Although some claim that Lindbergh was hoodwinked by the Germans into vastly overrating their capabilities, his reports back to the Air Corps, and subsequent public statements, did much to underscore the inadequate preparations of the U.S. Military, especially in aviation, for world conflict.

In addition to this confusing maelstrom of conflicting opinion, General Hap Arnold, commander of the U.S. Army Air Corps was actively campaigning against the export of U.S. built aircraft to Great Britain. General Arnold rightfully feared that any aircraft sent abroad to foreign powers would come at the expense of the Air Corps, a service which desperately needed to increase its own inventory.

The U.S. military and defense industries have most often been represented as being woefully unprepared before Pearl Harbor. However, in light of the above mentioned facts, and due in large part to the War in

As the courtroom filled up, additional bystanders gathered outside the courthouse, and across the street at the Oilmen's & Merchant's National Bank to discuss the case and await the outcome of the trial.

hadn't got stuck behind a coal truck just the other side of Fairchance."

"This is a travesty!" shouted Blakeley, standing once again. "There wasn't any coal truck!"

The Magistrate banged the gavel, more forcibly this time. "You there!" he said, pointing the gavel at Blakeley. "You've already as much as admitted that you were doing sixty. What's another twenty miles an hour? And you, Trooper -"

He searched the paper before him through his thick pince-nez glasses. "Trooper Corbett, is it?"

"Yes, your honor," said the trooper.

"Trooper Corbett, control your prisoner. One more outbreak like that in this courtroom, and you can take your next speeding ticket to the magis-

Although not a fighter, but rather a tactical attack bomber, the Brewster XA-32 is noteworthy, as it is one the few Brewster aircraft to enter development following the demise of the Brewster "Buffalo." Grossly overweight, the XA-32 was as heavy as many contemporary twin engined bombers. Its sole claim to fame seems to be that in its performance trials, it failed to meet any of the required specifications. Both models were scrapped following the trials, and the company went bankrupt shortly thereafter. (Photographs courtesy of the U.S. Air Force Museum)

trate court up in Connellsville. If I had a bailiff, I'd already have thrown him in jail for contempt!"

Blakeley sat down, and visibly attempted to compose himself.

"I'm sorry, your honor." Blakeley said. In the back of the courtroom, the second courtroom loafer returned, followed by six former patrons of Milligan's Bar. They took seats in the back of the courtroom.

"As long as you really mean it," said Magistrate Cockthorn. He was trying his best to be magnanimous, but his lower lip still puffed out. "And how does the culprit, I mean the alleged culprit, plead?"

"Excuse me, your honor," Trooper Corbett jumped in before Blakeley had a chance to respond.

Europe, the American aviation industrial establishment was actively engaged in the attempt, at least, to develop advanced fighter aircraft.

Some of these proved more successful than others.

The pre-war rush to develop new pursuit models that could hold their own against the new European designs would not bear fruit until later in the war; although U.S. aircraft fighters would ultimately achieve supreme air dominance, the desperate search for alternatives to the existing inventory of Army Air Corps pursuit models led Blakeley on his fateful mission.

"I'm afraid there's a little more."

Magistrate Warren T. Cockthorn glared at Blakeley.

"What do you mean, a little more?" he asked ominously.

"Well, there's more to it than just speeding," said the trooper. "I haven't had the chance to look up the statutes, but I'm pretty sure these other actions of his are against the law."

Blakeley sat fuming on the wooden bench.

Magistrate Warren T. Cockthorn's already beady eyes narrowed significantly behind his pince-nez. "Such as?"

"Well, your honor," said the trooper, "if he's not an army officer, and I have reason to believe he isn't, then he's impersonating an officer. Then there's creating a public disturbance, breech of the peace, inciting to

riot, I'm sure there will be more when I get a chance to look them up. But that's not the worst of it."

There was a scattered murmuring from the group of spectators in the back, which had now grown to a dozen or so. A bottle was being passed surreptitiously between the Milligan expatriates. Two people on the benches nearest the aisle got up and hurried out of the courtroom door.

"Order in the court!" Magistrate Warren T. Cockthorn yelled at the figures in the back of the courtroom. The bottle was quickly hidden. Then, Cockthorn's chin jutted forward, his face set in a grimace normally reserved for the worst kind of baby killer. He stared Blakeley squarely in the eye. "Not the worst, eh?"

"No, sir, your honor," said Trooper Corbett. "I have good reason to believe that this alleged culprit

September 13, 1939: Magistrate Warren T. Cockthorn's courtroom in more peaceful times. This photograph of an earlier case eloquently showcases the courtroom in all of its decorum. Justice Cockthorn presides, flanked by two American flags. In the center aisle sits Stella Periwinkle, the court stenographer. (File photograph courtesy of The Uniontown Citizen-Advertiser, SIdney Welt, photographer)

is an enemy agent."

"This is an outrage!" Blakeley jumped to his feet.

There was pandemonium in the courtroom, as spectators began to flock in through the courtroom door in droves.

"Order in the court!" yelled Cockthorn, banging his gavel repeatedly. "The defendant will have ample time to answer the charges. Please, continue, Trooper."

"I have reason to believe that this man, whatever his real identity, is impersonating an Army officer for the purpose of proceeding westward, to contact a cell of German or Japanese agents, possibly both. He has claimed that these German and Japanese terrorist cells are proceeding in this direction, attacking both military and civilian targets along the way. Our entire country may be in danger from the actions of this . . . . alleged culprit."

"Ummmppphhhh!" said Blakeley, barely able to restrain himself. But with one withering look from the Magistrate, he fell silent.

"Is that all, Trooper Corbett?" asked Magistrate Cockthorn.

"Isn't that enough, your honor?" asked Corbett.

The crowd erupted. Cockthorn banged his gavel futilely.

"Ummmppphhh!" said Blakeley.

"Order in this courtroom!" screamed Magistrate Cockthorn. "I'll have order, or I'll clear the room!"

LT. BLAKELEY COPS A PLEA

People were pouring into the courtroom now; men, women, and children. The benches were rapidly filling up. Arnold Snookett, just out of journalism school, and the sole reporter for the Uniontown Citizen-Advertiser, slid in behind Blakeley. He wore an Argyll vest sweater and a ratty brown snap-brim fedora. He pulled a pencil out of his hatband, whetted the tip between his lips, flipped open a steno notebook and began taking notes.

Sidney Welt, the photographer for the Citizen-Advertiser, nattily dressed in white bucks, tweed pleated pants and a loud plaid sports jacket, slid in beside the reporter. He was carrying a giant Speed Graphic camera with a huge round photoflash. The photographer sat down his Speed Graphic, and removed his snappy new plaid sports coat, draping it

Arnold Snookett stands behind Sidney Welt at the 1948 convention of the Pennsylvania Association of Journalists and Press Photographers. Having moved up in the world, (both Snookett and Welt were by then employed by the Pittsburgh Police Recorder), Snookett had relinquished his collegiate garb, and Welt now favored conservative pin striped suits. (Photograph courtesy of the Pennsylvania Association of Journalists and Press Photographers.)

neatly across the back of the bench. He settled into his seat with a flourish. When he was finally settled, he reached into his photo bag and pulled out a film holder, which he noisily slid into the back of the camera. He pulled out the slide and stashed it back into his photo bag.

"I'm ready, Arnold!" Sidney yelled to the reporter. Blakeley turned to glare at the photographer. But, looking out in the courtroom behind him for the first time, his eyes opened wide in wonder. The courtroom was almost full.

Magistrate Warren T. Cockthorn smiled at the photographer. He opened a drawer in the magistrate's bench and pulled out a sheaf of paper, which he held out to the State Trooper.

"Trooper," said Cockthorn, "would you hand these papers to the esteemed gentlemen of the press seated behind you?"

Trooper Corbett, with a look of uncertainty, took the papers, and handed them to the reporter. The reporter looked at the papers, and then looked questioningly up at the Magistrate.

While Lieutenant Blakeley dallied in Pennsylvania, the Axis war machine continued to hammer away at the Allies, fighting simultaneous air battles over Europe, North Africa, and with the U.S.S.R. on the eastern front. One can easily understand the Lieutenant's consternation with the legal hindrance to his mission. (Photographs courtesy of the U.S. Air Force Museum and the U.S. Navy)

"My press kit," said Magistrate Cockthorn, giving the reporter his best election winning grin. "Just so you can get my name right."

Arnold the reporter smiled, and gave the judge a thumbs-up sign. At the same time, he poked the photographer in the ribs.

"Owww, Arnold!" yelled Sidney. "That hurts!"

The Magistrate's smile disappeared as quickly as it came. Stone faced, he turned his attention once again to Blakeley.

"And how does the defendant plead?" asked the Magistrate.

"Your honor," said Blakeley, stepping forward. "If you would just-"

"I SAID- -How does the defendant plead?!"

"Not guilty, your honor," said Blakeley glumly. Deflated, he sat down on the bench. A raucous murmur passed through the crowd.

"Hmph!" said the Magistrate. "Well, now that we have that out of the way, does the alleged culprit have any evidence to present in his defense?"

"Your honor," said Blakeley, standing. "My orders and files are in these folders here. If you would just look at them, I'm sure that you would see that this is all just a big misunderstanding."

"I, being the judge, will be the best judge of that," said Cockthorn, officiously, rolling back in his swivel chair and hooking his thumbs in his suspenders, playing to the press and the crowd. "The alleged culprit may approach the bench with the alleged evidence."

Blakeley, barely keeping his slow burn under control, stepped up to the rickety table and threw down the file folders. The crowd oohed and aahed.

"What's all this?" The Magistrate indignantly flipped at the corner of the files.

"As I said, these are my orders and all the files germane to my mission," Blakeley said, doing his best to keep his voice level. "Your honor, the material in these files is of a highly sensitive nature. I shouldn't be showing them to you, even as a court official. But if it will help resolve this matter, I feel it is in the national interest that you view the appropriate documents. However under no circumstances should these files be revealed to the public. The nation's security would be at grave risk."

The crowd booed loudly. The Magistrate banged his gavel, and the booing subsided somewhat.

"Well, if anything in there is pertinent to this trial, now would be a good time for you to pull it out of your folder and present it to the bench. I'm not here to waste the taxpayer's money by doing your legal research for you!" The Magistrate banged his gavel loudly on the bench. He banged his gavel once more, even louder, looking expectantly at Sidney the photographer. Sidney looked dumfounded. Arnold the reporter poked him in the ribs again.

"Owww, Arnold!" yelled Sidney.

"Gwan!" hissed Arnold. "Take his picture, stoopid!"

While Sidney Welt remained with the Citizen-Advertiser throughout the war years, Arnold Snookett volunteered for a tour of duty as a waist gunner on B-24 Liberator bombers, where his walleye gave him a singular advantage in leading his aim at attacking German fighters. (Photograph courtesy of the Office of War Information)

Magistrate Warren T. Cockthorn arched his head and, smiling broadly, presented his best profile to the photographer, all the while vigorously banging his gavel on the bench. Sidney snapped the shot. The magnesium flash on the big Speed Graphic lit up the entire front of the courtroom for a fraction of a second. Blakeley's hands flew up to his face, almost as a reflex.

A rustle of awe passed through the crowd. Sidney noisily pulled the slide from his photo bag and slid it back into the camera with a loud clack. He pulled the film holder out of his camera, and flipping it over, rammed it back into the camera, readying the Speed Graphic for the next shot. Then, he pushed the flashbulb release with his right hand. The used bulb popped into the air in a short arc and Sidney deftly caught it in his left hand.

"Owww, Arnold!" he yelled as the smoking bulb burnt into his hand.

"Don't yell at me!" said Arnold. "I didn't do nothin!"

Sidney threw the bulb up in the air. It fell beside Trooper Corbett, exploding with a loud pop. The trooper jumped and turned, drawing his Smith & Wesson Police Special in a single slick motion. The crowd applauded vigorously.

Sidney gave Trooper Corbett a sheepish grin as he pointed down to the flashbulb pieces scattered on the wooden floor around the trooper's motorcycle boots. Trooper Corbett looked down, scowled at the photographer, holstered his gun, and sat down. The crowd applauded once again.

"Order in the court!" bellowed the Magistrate, banging his gavel feverishly. "Order in the court!"

Slowly the applause subsided. In the distance could be heard the sound of a siren. More people were now flowing into the courtroom. They overflowed into the back of the galley, as there was now standing room only in the crowded courthouse.

Magistrate Cockthorn picked up the folder and slid it across the bench to Blakeley. Outside, the siren was growing louder.

"Pick out your best stuff, soldier boy!" said Magistrate Cockthorn. "And make it snappy!'

As Blakeley took the file and sorted through it, the siren sound grew louder, until it filled the entire courtroom. Then, abruptly, it began to wind down as the siren was cut off. The side door to the courtroom opened, and Sheriff Eustis Poteet walked across the front of the courtroom. He walked up to the Magistrate's bench and addressed Cockthorn.

"Afternoon, Warren," said the Sheriff. "Heard there was a bit of a ruckus at the court house, so I just came by to see if you needed a hand."

"Everything's under control for now, Eustis," said the Magistrate. "Is Deputy Poteet with you?"

"He's out back in the patrol truck," said Poteet. "Want me to get him?"

"Just stand by," said Magistrate Cockthorn. "You never know when these things could get out of hand."

"Sure thing," said the Sheriff. He looked around at Blakeley, glaring menacingly. "What is he, Warren, one of those deserters?"

"Enemy agent," said Cockthorn.

"Only thing worse than a deserter!" the Sheriff growled, his thin lips pulled back to reveal his teeth in a grimace of hate. "We'll be outside in the squad car if you need us."

Sheriff Poteet crossed the courtroom, turning his head to keep a watchful eye on Blakeley until he was out the side door.

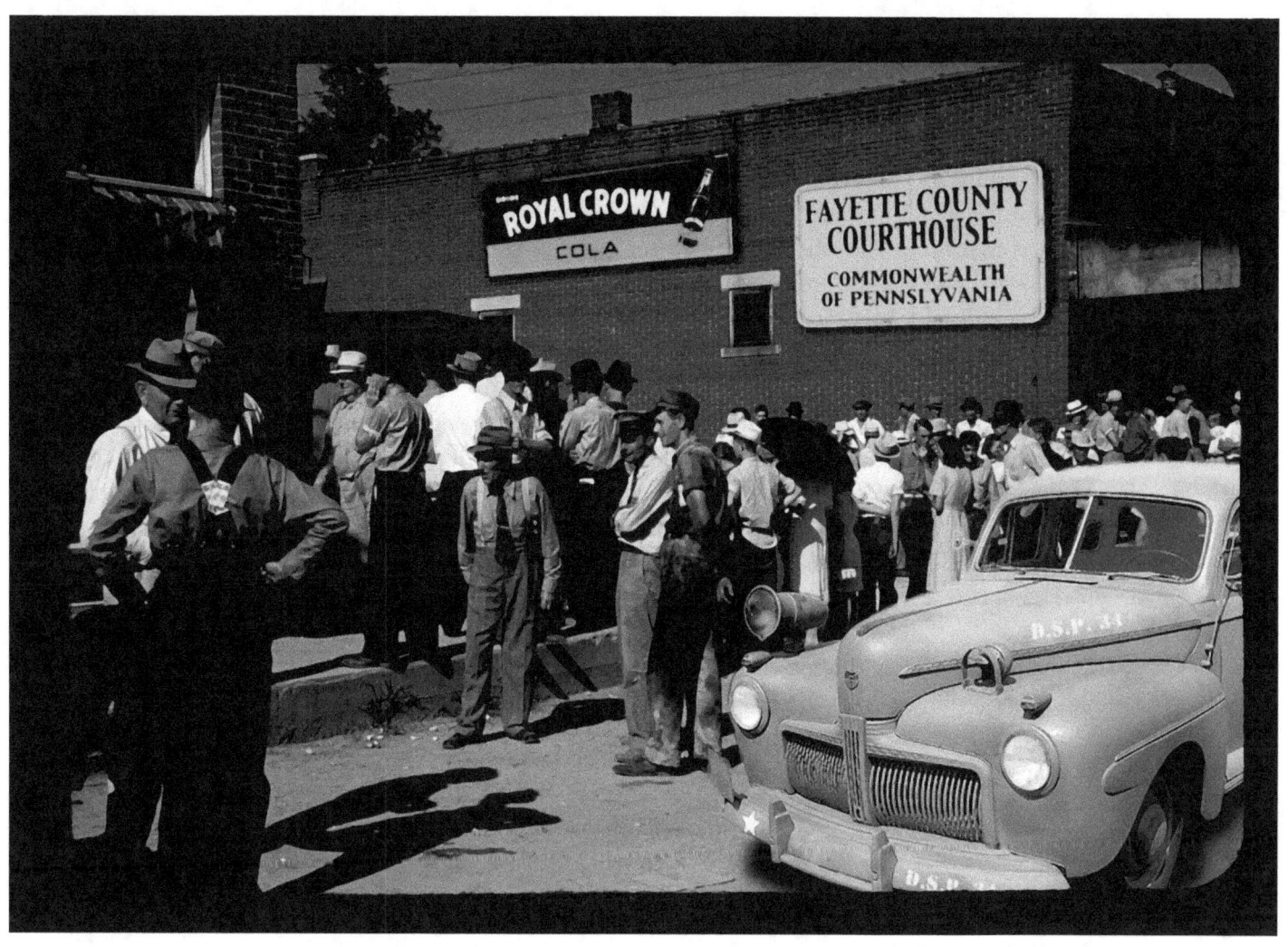

The courthouse was packed, and the crowd overflowed into the alley beside the courthouse, awaiting news of the trial's outcome. Lt. Blakeley's 1942 Ford staff car is to the right.

5. JUSTICE IS SERVED:
THE GERMAN TERRORISTS ARE UNMASKED

(Excerpt from *The Flight of the Boomerang*, by Elmer C. Wackmallit, continued.)

Fayette County Courthouse
1212 North Court Street
Uniontown, Pennsylvania
August 23, 1942 3:23 p. m.

Blakeley, having pulled three sheets of paper out of his file, handed those papers to the Magistrate. Cockthorn took the papers, adjusted his glasses and read the first page. He held the first of the papers before Blakeley.

"This paper here, what's this got to do with anything?" Cockthorn asked petulantly.

"Those are my orders directing me to my destination," said Blakeley.

"So you admit you were enroute to Council Bluffs, Iowa?" asked the Magistrate.

Trooper Corbett jumped up, turning to face the crowd. Sidney the photographer dutifully snapped his picture.

"That's where the Jap and Kraut terrorists are coming from!" he yelled.

The crowd erupted in a roar. Arnold the reporter was writing frantically. Sidney hastily changed the film holder for another one in his photo bag.

"Your honor, I object!" yelled Blakeley. "Those orders are secret. To reveal them openly in this courtroom directly contradicts the national interest!"

"Your interest, you mean!" said the trooper. The crowd roared in approval again.

"Mr. Blakeley, if that is indeed your real name, this case cannot be tried without the presentation of evidence, and I will decide what evidence is appropriate to present in my courtroom!" Cockthorn struck the gavel with all the force of his conviction. The handle broke, the gavel head twirled up in the air and hit the whirling fan blade. The blade knocked the gavel head back down into the courtroom spectators, where it struck the blond head of a huge woman in a gaudy yellow and pink check summer dress who was sitting on the edge of a courtroom bench, her purse in her lap.

"Oooooh!" moaned the woman. She stood up, and, still holding her purse, twirled completely around once and fell into the aisle with a tremendous thud. The entire courthouse reverberated with the blow. The crowd roared in approval.

The woman's gargantuan legs flew up in the air and crashed down, flipping up the woman's dress, revealing her calf-length rolled up stockings, and just the scantiest view of the bottoms of her bloomers.

"Myrtle!" yelled a skinny man with a pencil mustache who was sitting beside her until just a few seconds before. He jumped up and ran to the woman, who now had a bump under her blond ringlets the size of a loon egg.

Sidney the photographer stood up on the bench, turned around and snapped a picture of the whole dreadful scene.

"Now, look here!" the skinny man shouted at Sidney. "You can't take her picture when she's exposed like that!"

The Magistrate threw the

broken handle over his shoulder. He picked up the papers and shuffled through them again, stopping with particular interest on one in particular.

Shortly, a man in coveralls and a carpenter's apron got up from his place in the third row, and stepped out to the aisle. He took a big three-pound carpenter's framing hammer from a holder on the side of his pants.

"Here, Warren," said the carpenter, holding the hammer out to the Magistrate. "You can use my hammer. Just don't break the handle."

"Warren, tell this photographer that he can't take a picture of Myrtle in the condition she's in!" yelled the skinny man.

Sidney popped out his flashbulb, but this time he caught it in his handkerchief. He held the flashbulb up menacingly at the skinny man, as if he was going to throw it. The skinny man retreated a few skittering paces up the aisle, out of range. The

Myrtle Canker in happier times before her horrific abduction. She is shown crossing the street from Adam's Drug Store after her daily banana split. Photograph courtesy of The Uniontown Citizen-Advertiser, from the May 1, 1939 feature story, "Uniontown Belles," by Arnold Snookett. Photograph by Sidney Welt.

Vernon Dudley proudly displays the hammer used by Magistrate Cockthorn to officiate at the trial of Lieutenant Julian Blakeley. (Photograph by Sidney Welt, courtesy of The Uniontown Citizen-Advertiser)

thing? I've got a trial to run here!"

Magistrate Cockthorn banged the hammer on the table, once, experimentally, then a second, and then a number of times in succession. He turned the hammer over once and held it up, looking at it appreciatively. Then he nodded in satisfaction to Mr. Dudley, who waved back from the galley.

"Your honor," Blakeley fumed, "This isn't a courtroom, it's a -"

"Mr. Blakeley!" boomed Magistrate Cockthorn, whacking the table with the big framing hammer, putting a sizable dent in the surface. "Need I remind you that only my excessive leniency has kept you from being in contempt of this court? Now give me a moment to review this document. In fact, I declare a ten-minute recess for me to read these documents and to remove the wounded from the courtroom."

At this, Homer and three hefty farm boys tried to pick up Myrtle, but they could barely budge her. Homer motioned to the crowd, and three more good sized men in farm coveralls joined in. Homer decided to supervise. He pulled down Myrtle's skirt, and picked up her pocketbook.

"One, two, three, heave!" shouted Homer. The straining men picked Myrtle up and carried her out the front door of the courthouse, with Homer leading the way.

"Where's your car, Homer?" the big man on the right front asked Homer as they reached the bottom of the courthouse steps. All six of them were panting heavily.

"It's about six blocks thataway," said Homer, pointing down the road. "It was as close as we could park, what with the big court turn-out and all!"

"Lordy!" said the big man. "She sure is a lot of woman!"

"No way we can carry her that far," said the man across from him, between huffs and puffs. "Let's lay her down right here, until we decide what to do with her."

carpenter took the hammer up to the Magistrate and returned to his seat.

"Thank you, Mr. Dudley," said the Magistrate. "The court appreciates your interest in seeing justice done. I'll see that you get your hammer back at the close of trial."

"And you, Homer," said Cockthorn, pointing the hammer at the skinny man, "On account of you being a barber, and not a legal man, you might not have heard of the first amendment, which guarantees the freedom of the press. So I'm going to overlook your outburst in my courtroom."

"Well," said Homer sulkily. "What are we going to do with Myrtle?"

"Get you some help and carry her out to your car," said Cockthorn. "Do I have to think of every-

"Boys!" cried Homer. "You can't just leave her here out on the street! You wouldn't do that to a defenseless woman, would you?"

"Well, we want to get back in the courtroom and see the rest of the show!" said one of the farm boys in the back, a hurt tone in his voice.

"Me, too!" said Homer. "But if she wakes up out here on the sidewalk, she'll surely be steamed up! You boys don't know what that woman's like when she gets mad."

"Think of something, quick!" said the big man. "My hands are slipping."

Homer looked around quickly. Then he spied Blakeley's Army Ford.

"I know," Homer said brightly. "Put her in that Army fella's car. He surely ain't going anywhere!"

"Yeah," laughed the farm boy. "Warren Cockthorn's going to jack up the courthouse and put him under it!"

Homer opened up the rear door of the Ford. With much straining and grunting, the six men finally got Myrtle into the back seat. Homer sat her pocketbook down beside her and closed the door.

"Whew!" said the farm boy. "I just hope she wakes up later, so she can unload herself."

"Let's get back inside quick," said the big man. "I don't want to miss any more of the show!"

Magistrate Cockthorn, still sitting at the Magistrate's bench during the recess, had been scanning the documents in silence. His eyes had grown wider with each line that he read. After a few moments, he gently laid the documents down on the table, drummed his fingers on the table, and started humming softly to himself.

Shortly, Homer and his crew reentered the courtroom. They took their seats as Cockthorn adjusted his pince-nez glasses, all the while smiling benignly at Blakeley.

"Mr. Blakeley, your documents," he said. He handed the papers to Blakeley, who put them back into his file.

There was a constant rumbling in the courtroom, as people rolled in after the recess and took their seats. Magistrate Cockthorn smiled serenely at the crowd as he tapped the bench lightly three times

Mr. Homer Canker proudly displays his patriotic window posters, little suspecting that he and his wife will soon become victims of Axis saboteurs. (Photograph by Sidney Welt, courtesy of The Uniontown Citizen-Advertiser)

with the big claw hammer.

"Order in the court," he said quietly, barely heard over the din in the courtroom. "Court is now in session."

There was immediate silence in the courtroom. Everyone present was anxious to hear the next development. The Magistrate gazed across the courtroom. He smiled at the reporter. He smiled at the photographer. He smiled at Trooper Corbett. He smiled at Blakeley.

"Mr. Blakeley," he asked mildly. "Who exactly is this Xontar the Magnificent? And just who are the Zenturions?"

"Your honor," protested Blakeley. "I've told you, these matters cannot be publicly discussed!"

"Mr. Blakeley, or whoever you are," said Magistrate Cockthorn, slowly building back up to his old

THE CONCEPT ALMOST TOO GOOD TO BE TRUE: THE TWIN ENGINE FIGHTER

The Z-44 "Boomerang" was initially conceived by Professor Hermann Flungk in the middle of the Great Depression. Although the thirties were a decade of economic hardship and privation, they were also a decade of tremendous social upheaval, creating conditions which would culminate in a war of unprecedented proportions. Within the aviation community, men of prescient vision were striving to develop the aerial weapons that would decide the coming conflict in the air. One of the solutions which seemed obvious at the time, but questionable in hindsight, was the concept of the twin engine pursuit aircraft.

Larry Bell was determined to be an innovative designer of military aircraft. Hence, the Bell YFM-1 Airacuda. Bell Aircraft Corporation's first fighter had a definite "Buck Rogers" flair, and surprisingly, there were no problems with the power transmission shafts which plagued most pusher designs. The concept was doomed for multiple reasons, and Bell went on to design the only slightly less exotic Bell P-39 Airacobra.

The Grumman XF5F "Skyrocket," above, and its sister ship, the Grumman XP-50, right, were co-developed to meet the needs of both the U.S. Navy and Army Air Corps for a high powered interceptor. However, neither aircraft saw service, joining the ranks of might-have-been fighters.

ADVANTAGES TO THE TWIN ENGINE CONCEPT

A consideration for safety has long been the hallmark of the multi-engined design. With a twin, obviously, if an engine fails, you have not lost 100% of your power as would be the case with a single engine aircraft. Safety, however, was not the primary reason for the intense interest in twin engine aircraft in the period immediately preceding World War II. The overriding concern was for power: power to lift heavy weapons systems, power to carry those weapons systems for long distances, and, above all, power for speed.

steam. "Of what possible significance to the national interest can it be that someone who calls himself 'Xontar the Magnificent,' eats tinfoil flavored cupcakes, thinks that he talks to beings from another planet, and seems to believe that he is going to construct an interplanetary spaceship from cardboard boxes and used tractor parts?"

Trooper Corbett jumped up. "It's code! That's what it is! Those supposed orders of his are faked in an obvious attempt to clear his passage to Council Bluffs, where he is supposed to deliver this message to this 'Xontar' character, whoever he is. All I know is, that name sounds German!"

A man jumped up in the third aisle. "Yeah! Xontar!" he shouted. "Just like Hitlar!"

Another man jumped up toward the back of the courtroom, "Yeah!" he shouted. "And that Zeppelin fella! Hindarberg!"

Another man in the middle of the courtroom jumped up on his bench. "Yeah!" he shouted, giving the Nazi salute. "Heil Xontar!"

The crowd erupted in a roar. People were jumping up on the benches. Someone else picked up the chant. "Heil Xontar! Heil Xontar! Heil Xontar! "

"These charges are preposterous!" Blakeley attempted to yell above the din. "You are interfering with an official United States Army Air Corps operation! I'll have you all hauled into Federal Court! I'll-"

But Lieutenant Blakeley was unable to complete his harangue. The rising vocal tide of the crowd overwhelmed his protestations. Then there was a tremendous crash as the rear doors of the courtroom banged open.

In the United States, the state of the art of aircraft engine development in the mid thirties was best, albeit broadly, represented by three engine manufacturers:

The Pratt & Whitney R-1830 Twin Wasp radial engine was installed on the Curtiss P-36-B, among others, developing 1100 horsepower.

The Wright R-1820 radial engine was installed on the Brewster F-2A Buffalo. It also developed 1100 horsepower in this configuration.

The Allison division of General Motors produced the Allison V-1710 liquid cooled series of inline V-12 engines, used in the Curtiss P-40 and the Bell P-39, among others. In the early configurations, these engines also produced approximately 1100 horsepower.

The German Messerschmitt Bf-110 ultimately proved too unwieldy for the role of daylight interceptor. Like the Bristol Beaufort, its primary fighter role was to be as a radar equipped night fighter.

Because power in a single engine aircraft is limited to the rated output of the engine (in this case, approximately 1100 horsepower by the mid-thirties), one way to overcome that limitation is to add additional engines. This simple fact led to a multitude of twin engine aircraft designs.

A MULTITUDE OF DESIGNS

One such design was the Bell YFM-1 Airacuda, first flown in 1937. This was the first fighter design produced by the Bell Aircraft Corporation. This was a totally unique concept by a corporation with no previous experience in military aircraft construction. The aircraft incorporated two Allison V-1710 engines driving pusher propellors, each mounted in a wing nacelle containing a forward firing gunner armed with a 20 mm cannon. As with all pusher configurations, engine overheating was a constant problem. The aircraft was envisioned as a "bomber destroyer," but, despite its sleek futuristic design and twin Allison engines, a cruise speed of 244 mph limited its ability to intercept its targets. Inherent design flaws, not the least of which was the difficulty of evacuating the crew in an aircraft with two pusher

Above: Bristol Beaufighter in U.S. markings, 416th Night Fighter Squadron, Grottaglie, Italy, 1943. (U.S. Air Force Museum) Inset: Beaufighter of No. 30 Squadron, Royal Australian Air Force, New Guinea. Sergeant W.B. Ball, photographer.

ALL RISE:
THE COMING OF THE BAILIFF

Bailiff O'Riley stumbled in through the courtroom doors. He pirouetted elaborately, nearly falling as he attempted to shut the door behind him. He stumbled up the aisle between the gallery benches.

"Bailiff!" shouted Magistrate Cockthorn, viciously banging the big hammer on the table. "Restrain this defendant! This . . . . insurrectionist! This. . . . vagabond! This. . . alleged culprit!"

Bailiff O'Riley, who now appeared to be slowly falling forward down the center aisle, rapidly increased his forward motion. Whether or not this was in response to Magistrate Cockthorn's command, was impossible to determine.

Upon reaching the front of the courtroom, Bailiff O'Riley's body cocked to the left slightly, as he vainly attempted to turn the corner and apprehend Blakeley.

Try as he might, he was unable to complete the maneuver. Instead, there was a look of surprise and consternation on the Bailiff's face, as if his body simply would not do what he was telling it to, and what in his heart he knew that it was capable of. O'Riley continued forward, his body half-canted, until he slammed full force through the balustrade. The court stenographer fell backwards, narrowly avoiding O'Riley, her spiked heels flying through the air. O'Riley then sailed over the heads of the astonished jury and across the magistrate's desk, driving

propellors, caused the demise of the type by 1942.

Another twin engine design was the Grumman XF5F "Skyrocket," developed from a 1938 Grumman proposal for a radical fighter design. The XF5F, powered by two Wright R-1820 engines exhibited superior performance to the Bell YFM-1 Airacuda, with a top speed of 383 mph. Still, issues of complexity with the design ultimately caused the U.S. Navy, the primary customer for the design, to opt for the much simpler single engine Grumman F4F "Wildcat."

The Lockheed P-38 was the aircraft flown by the two top scoring U.S. aces, Major Richard Bong and Major Thomas B. McGuire, Jr. Shown here is a YP-38, one of the initial service test P-38s. Inset: The cockpit of the P-38, containing a control yoke rather than a stick, which was very unconventional for a fighter.(Photographs courtesy of the U.S. Air Force Museum)

Above, a P-61A of 419th Night Fighter Squadron. The first aircraft to be designed as a night fighter, and a radar equipped interceptor, the P-61 was larger than some contemporary medium bombers. The last Allied air victory of the war was achieved by a P-61 just before VJ day. Inset: A P-61 of the 6th Night Fighter Squadron in the Mariana Islands in late 1944. Note the radar visible beneath the translucent radome, backlighted by the setting sun. (Photographs courtesy of the U.S. Air Force Museum)

Foreign twin engine fighter designs were also developed during this period, most notably the German Messerschmitt BF-110, and the British Bristol Type 156 "Beaufighter." It is noteworthy that both of these aircraft achieved their greatest success as night interceptors fitted with radar, usually in a defensive role against attacking bombers.

In short, the promising concept of the twin engine fighter, based largely upon the need for more power, did not bear the intended fruit. Twice the engines meant twice the fuel consumption, increased drag from the engine frontal area, increased crew, extended training periods, and a

his head into the astonished Magistrate's solar plexus.

"Heil Xontar!" the crowd chanted relentlessly. "Heil Xontar! Heil Xontar!"

The Bailiff's momentum carried him onward, into the lap of the Magistrate. Cockthorn's swivel chair sprawled backward, His legs, flailing out reflexively in an effort to keep him upright, kicked over his most recent tomato can spittoon. The first spittoon knocked over the next of the tomato can collection, and then, the next, like a cascade of falling dominoes.

A tidal wave of viscous brown slime careened across the courtroom floor. Cockthorn and the Bailiff crashed against the rear wall of the courthouse, the Magistrate's pince-nez glasses falling to the floor. Cockthorn's arms flew out as he impacted the wall, under the full weight of the Bailiff. The big three-pound framing hammer flew up into the air, out of the Magistrate's hand.

Simultaneously with this action, Sidney the photographer snapped the action with his big Speed-graphic. The chanting started to break up, as confusion spread through the crowd. Sidney, anxiously preparing for the next shot, ejected the bulb, which crashed into the floor, where it exploded, sounding just like a gunshot.

The big framing hammer crashed with a resounding bang through the courthouse window, spraying a shower of glass across the alley. It continued through the

marked decrease in maneuverability. Additionally, as the war progressed, engine power increased dramatically, and the need for more than one engine was relegated to bombers and transports.

Of course, there were notable exceptions. The Lockheed P-38 "Lightning" was built in response to a 1937 Army Air Corps specification. Powered by two Allison V-1710 engines, it achieved a top speed in excess of 400 mph, becoming one of the best fighter aircraft of the war. Other twin engine fighter aircraft

If the Luftwaffe had possessed the Dornier Do-335 and the Messerschmitt Me-262 in any numbers, the outcome of the Air War in Europe would have been very different. (Do-335 image courtesy of Ad Meskens. Me-262 photograph courtesy of the U.S. Air Force Museum)

worthy of note include the Northrop P-61 "Black Widow," the first U.S. aircraft specifically built as a night fighter, and two German designs, the Dornier Do-335 "Arrow," and the Messerschmitt Me-262. Although the Do-335 was not developed until later in the war, it is noteworthy for its unusual center-thrust design. This drag reducing feature contributed to the Do-335's amazing top speed of 474 mph. The Me-262 is especially noteworthy as the first operational jet fighter. When it entered combat in the European Theater in the closing stages of the war, it presented

FLUNGK Z-44 "BOOMERANG"

Flungk Mark XIII "Superblow," as installed at the pilot's station of the Z-44 "Boomerang."

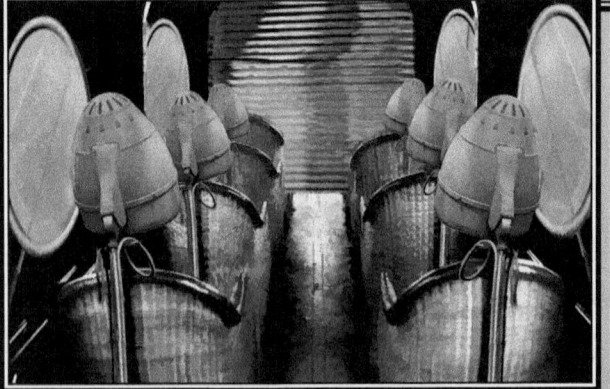

Although the Flungk Z-44 "Boomerang" was not initially noteworthy for any of the attributes of contemporary Pursuit or Intercept aircraft, it nonetheless achieved limited notoriety as the world's first aerial hair salon. (Images courtesy of the Whitley Speale Collection of the Bone Lake Research Museum)

a serious threat to Allied bombers.

Last, and quite possibly least, it would be prudent to mention the Flungk Z-44 "Boomerang," an aircraft of such striking originality and unique flight characteristics, that only the fact that it had two engines could possibly warrant its inclusion in this august group of aviation thoroughbreds. The convoluted and often misconstrued history of this arcane aircraft has often been neglected by most competent aviation historians. Indeed, the story of how the Z-44 entered the pantheon of legendary World War II fighter aircraft is much too complicated to elucidate within the confines of this short treatise, and will be elaborated on at length within the remaining pages.

side window of the 1937 Chevrolet patrol pick-up, spewing an additional shower of glass on Deputy Poteet and Sheriff Eustis, who were dozing in the front seat.

In the courtroom, Bailiff O'Riley slid off of Magistrate Cockthorn, coming to rest in the floor. The Magistrate's chair rebounded forward, and one wheel of the office chair rolled over his pince-nez glasses with a sickening crunch. Cockthorn wheezed, gasping for air, his breath knocked out of him.

Behind Blakeley and Trooper Corbett, Sidney Welt, the photographer, pulled out another film holder from his photo bag and slapped it expertly into his Speedgraphic. The solid click of the film holder sliding home sounded to Trooper Corbett exactly like a gun being cocked behind him.

Simultaneously, out in the alley, Deputy Poteet drew his five-shot top-break Harrington & Richardson pistol from his holster and fired a warning shot through the roof of the patrol truck.

Inside the courthouse, Trooper Corbett pulled his K-frame thirty-eight caliber Smith & Wesson Police Special and spun around, grimly searching the immediate area for threats of a criminal nature.

Back out in the alley, Eustis, curious as to why the truck wouldn't start, finally figured out that he didn't have the key on. He turned the ignition to hot, and the souped up Chevy flathead engine roared into life with a tremendous backfire that blew the muffler off. The engine quickly revved to the redline, it's fan belts screaming.

Inside, unable to locate a specific target, Trooper Corbett fired his pistol at the ceiling four times in a vain attempt to restore some semblance of order. Two of his bullets struck the ancient fan, which fell, blades still revolving, directly into the panic stricken crowd in the aisle.

With the addition of the fallen fan, the courthouse crowd, already confused, transitioned into a panic, which instantly notched up to mass hysteria.

People in the front of the courtroom were now screaming. Those knocked down by the fan were trampled underfoot as the mob pressed to the exits. Sidney lifted his Speed Graphic, hastily inserted a fresh flashbulb, and took a picture of Trooper Corbett, smoking gun in the air. The trooper promptly slammed his left fist into the photographer's jaw. Sidney slumped to the floor. Arnold the reporter slowly stood up, raising his hands over his head in surrender.

6. LIKE UNTO THE TRIALS OF JOB

Above: The grim aftermath. Certain proof of the work of Axis saboteurs. (Photograph courtesy of The Uniontown Citizen-Advertiser. Photograph by Sidney Welt)

THE TRIBULATIONS OF EUSTIS AND DEPUTY POTEET

(Excerpt from *The Flight of the Boomerang*, by Elmer C. Wackmallit, continued.)

Outside in the alley, Sheriff Eustis Poteet discovered that he had neglected to take the Chevy patrol truck out of gear before starting it. The patrol truck, engine screaming, sped out of control down the alley. The careening vehicle bounced off the brick wall of the courthouse with a scraping sound almost exactly like fingernails on a blackboard, accompanied by the sickening crunch of bending fenders. Eustis fought with the flailing steering wheel in a heroic effort to regain control of the elusive Chevy. In an instinctive reaction, he pressed down even harder with his left foot, which under normal circumstances would be on the brake. However, this foot was now on the gas pedal, and, to Eustis' consternation, instead of screeching to a stop, the souped-up Chevy flathead screamed like it was going to fly into a million tiny pieces.

The patrol truck lurched off the courthouse wall and veered across the narrow alley, impacting the brick wall of the drug store with an even higher pitched scrape and crunch. The jaunty little pheasant radiator cap shot straight up in the air as a billow of oily steam erupted, enveloping the Chevy's windshield in a greasy vaporous fog. Deputy Poteet, sensing imminent danger and eager to do his part, fired another warning shot, this time through the front windshield, sending spider web cracks running through the glass. Eustis finally got a tenuous hold on the spinning steering wheel and more or less guided the spoke-wheeled missile down the center of the alley. He pushed as hard as he could with his left foot, but instead of slowing down, the sassy little patrol truck, its siren screaming and its engine on the verge of self-destruction, continued to accelerate, heading straight for Court Street.

Inside the courthouse, Bailiff O'Riley finally came to amid the manic din of the crowd, which by now was completely out of control. Vaguely, he registered what sounded like multiple gunshots from at least three different weapons. With a heroic effort he pulled himself up. He peered across the courtroom.

Through bleary eyes he saw a figure with a gun crack the head of someone else, who was holding

Sheriff Eustis Poteet discusses post-trial strategy with an unidentified party in the alley between the courthouse and Adams Drug & Assorted Sundries. The unidentified party is most likely Deputy Poteet. (Photograph courtesy of The Uniontown Citizen-Advertiser. Photograph by Sidney Welt)

what looked like a big camera. In his advanced state of inebriation, it was impossible for Bailiff O'Riley to discern that the striker was indeed a brother lawman, in fact, Trooper Corbett.

All O'Riley knew was that the photographer, and the little reporter beside him who was now holding

up his hands, were good guys. They had even bought him numerous drinks in the past at Milligan's Bar. It didn't matter to O'Riley that they were just trying to pump him for some courthouse gossip. They were his drinking buddies, and now someone was whacking one of them on the head and holding the other at gunpoint! In a tearful and slobbering rage of drunken melancholy, Bailiff O'Riley charged at the trooper, screaming incoherently. He was incoherent, because in his present state, he couldn't remember the names of either of his news buddies.

There was just enough room on the dais for Bailiff O'Riley to build up a good head of righteously indignant steam before he stepped into the spreading contents of the tipped-over tomato cans. He plopped with a viscous smack into the oozing brown goo. His forward momentum carried his screaming body across the dais, over the edge, and into the back of Trooper Corbett, where he landed with both size thirteen boots, knocking the trooper out cold.

The unfortunate trooper, having absorbed the Bailiff's kinetic energy, slowed O'Riley's forward progress somewhat. Miraculously, the Bailiff came to rest upright. He teetered there for an extended instant, like a giant oak, which, having been cut, cannot decide whether to fall into the notch, with the wind, or just say screw it all and fall downhill.

Out on Court Street, even in these days of gas rationing, there was a moderate amount of traffic. This was the county seat, after all, and there was always official business to be taken care of. In addition, with the war on, the oil fields were running full bore. Construction equipment and oil rig trucks constantly cruised northbound into the oil-fields, and tanker trucks were constantly coming southbound, heading for Route Forty, then up to Pittsburgh and the Ohio River ports.

It was now late in the day, and it was the closest that Court Street ever came to having a traffic jam. There were a number of cars and lumbering clumsy trucks, but they were still moving at a fast pace.

It was into this relatively cluttered mass of vehicles, that the Sheriff Eustis Poteet's patrol truck, with its siren screaming a useless warning of impending doom, its fenders flapping in the wind like the feathers on a war bonnet, its fan belts screeching like the banshees of a thousand lost souls, now came shrieking out of

A Keystone Oil Company tanker truck beneath an overpass in Pittsburgh. It is possible, though unlikely, that this is the very truck involved in the traffic altercation with Sheriff Eustis Porter's patrol truck. (Photograph from the morgue files of the Pittsburgh Police Recorder, June, 1941)

The pigs began to squeal frantically as the Model T, in which only an instant ago, Homer Wells was tooling down Court Street minding his own business, was now pointed directly at the front of Adam's Drug and Assorted Sundries. The Model T farm truck slammed through the facade of the Drug Store, busting up the check-out counter, and, after plowing through a display of Whitman candies, finally came to rest in the incontinence section. The rear doors of the hog hauler flew open, spilling berserk porkers into the premises of the drug store.

Blakeley came to. At first, everything was gray. The gray slowly dissolved into little spots before his eyes. He was finally able to suck in a couple of breaths of air and his vision finally cleared. It was only then that he noticed a warm and slimy substance, which reminded him of egg whites, slowly soaking through his uniform, coating his chest and legs. With the superhuman strength that only comes with true disgust, Blakeley threw the Bailiff off to one side. Slowly, he stood, arms akimbo, looking with dismay at the river of tobacco juice slowly streaming down the front of his uniform.

Sheriff Eustis Poteet's patrol truck continued across into the southbound lane, directly into the path of an oncoming Keystone Oil semi-truck tanker. The driver whipped the wheel of the red Peterbilt tractor to the left, in a vain attempt to pass behind the flying Chevy. Forward momentum, though, won out against the driver's intentions. The truck and trailer rolled over sideways. The big red trailer burst open. Court Street was instantly flooded with high grade Pennsylvania crude.

The sliding trailer just missed the tail end of the Chevy patrol truck. In the car, Eustis held the wheel in a death embrace. Deputy Poteet stared straight

the courthouse alley like an unguided missile.

In the courtroom, Bailiff O'Riley's eyes rolled back in his head, his arms flailed outward in an automatic attempt at balance that must have originated very low in his reptile brain. His hands fluttered tremulously at the end of his wrists like the delicate fans of a geisha. He turned such a sickly green pallor that, for a moment, Blakeley had an unwanted image of a giant effeminate Frankenstein monster attempting to dance ballet. Then O'Riley toppled forward, aimed squarely at the transfixed and utterly speechless Blakeley. At the last instant, the Bailiff's arms flew out and embraced Blakeley in what could only have been an automatic reflex to break his fall. Blakeley fell to the floor, beneath the tobacco juice soaked Bailiff. He quickly lost consciousness as his breath was knocked out of him.

The little Chevy pick-up barreled into the flow of traffic on Court Street with an abandon that was normally reserved for the truly insane. Crossing into the northbound lane with absolutely no regard for the sanctity of human life, the patrol truck clipped the tail end of Homer Wells' ancient Model T, which had been cleverly converted into a hog hauler. The back of the Wells truck, as usual, was loaded with pigs.

A PAGE OUT OF HISTORY:
THE UNIONTOWN CITIZEN-ADVERTISER, AUGUST 22, 1942

(Editor's note: The discovery of the newspaper fragment shown on the following page prompted noted Fayette County Historian and Auctioneer Scott Thomas Neucomber to research the morgue files of the defunct Uniontown Citizen-Advertiser for the complete story and photographs on Blakeley's day in court. That story, as it originally appeared on page 5 of the August 22, 1942 edition, along with the photographs as found in the morgue files, appears here in its entirety, courtesy of Mr. Neucomber.)

VIOLENCE ERUPTS AT COURTHOUSE DURING ENEMY SPY TRIAL

The War came close to home yesterday, striking with vicious violence at the Fayette County Courthouse yesterday afternoon. State Trooper Patrick Corbett had apprehended a suspect earlier in the day, whom he had brought before County Magistrate Warren T. Cockthorn. The suspect had been stopped for a routine

Trooper Corbett recuperates on the courthouse lawn following the altercation with the "Nazi attackers." (Photograph courtesy of The Uniontown Citizen-Advertiser. Photograph by Sidney Welt)

ahead, pistol ready. They were both screaming, but that seemed to have no effect on the performance of their vehicle.

In the courtroom, Magistrate Cockthorn, sprawled against the back wall, twitched, but didn't stir. Blakeley kicked at the supine form of the Bailiff with loathsome revulsion. The screaming crowd, further incited by the caterwauling pandemonium in the street outside the Courthouse, seemed convinced that the entire city was now under attack from the Japo-German terrorists, or that the world was coming to an end, whichever event was the most disastrous.

People near the outside walls of the courtroom had by now stopped trying to fight their way into the aisle, and were instead busting out the windows, sometimes by hurling a fellow citizen through one. Then they climbed out into the courtyard across the broken glass.

Trooper Corbett remained in a state of apparently blissful unconsciousness.

UNBENT ALLEGIANCE TO A HIGHER AUTHORITY:
LT. BLAKELEY GOES UNDERCOVER

Blakeley looked around warily. It seemed that no one was evidently concerned about his whereabouts, or his condition of confinement. Quickly, he took off his campaign jacket and turned it inside out. Using the jacket as a rag, he wiped off as much of the gooey tobacco juice as he could from the front of his trousers and shirt. Blakeley spied the photographer's plaid jacket lying across the back of the bench behind him. He picked it up and put it on. It was a little tight across the shoulders, and the sleeves were a tad short, but it would do.

Blakeley gazed out again across the crowd. They were pushing toward the exit in the back of the courtroom so hard that the people in front were jammed against the doors and unable to move. He looked to his left, eyeing the door where the constable had taken his prisoner out earlier. Slowly at first, he moved toward the door. No one seemed to take notice, so he walked casually over to the door, opened it and looked in.

There was a small holding cell on one side of

the anteroom and a couple of wooden chairs on the other side. The chairs were pulled up to a porcelain topped table. A ratty deck of playing cards was centered in the middle of the table. The constable was nowhere in sight, obviously taking advantage of his position by the door to beat a hasty retreat when the festivities were initiated. In the cell, the constable's prisoner was sprawled on a bunk, his red checkered shirt pulled up over his head, snoring soundly. Blakeley took one last look back into the courtroom. The reporter was still sitting on the bench, hands in the air, staring at Blakeley. Upon seeing Blakeley looking at him, he picked up his steno pad and started writing feverishly.

LUTHER HANKUS SAVES THE DAY

Meanwhile, the screaming Chevy patrol truck smashed between a beautiful baby blue Model A Ford coupe and a Dodge Brothers town wagon that were parked across the street from the courthouse. The left wheel and axle of the Chevy pick-up took off the hood and radiator of the Ford, while the wooden structure of the Dodge town wagon was busted into toothpicks and shattered glass.

The Sheriff's Chevy continued across the sidewalk, busting through the revolving glass door of the Oilmen's and Merchant's National Bank. The door imploded in a spuming shower of glass that reached all the way back to the door of the President's office, where it came gently to rest like a sparkling ocean wave.

The little pick-up was squeezed between the substantial stone pillars on each side of the revolving

A remnant of the front page of the Uniontown Citizen-Advertiser for August 22, 1942, which had been used as wrapping paper in a box of depression era glassware. It was serendipitously discovered by an alert auctioneer during an estate sale, and saved from oblivion.

traffic violation on State Route 40, but Trooper Corbett became suspicious when the suspect began rambling about a secret meeting with enemy terrorist agents in the Council Bluffs, Iowa area.

"He was dressed in a U.S. Army uniform," stated Trooper Corbett. "He was even driving a Ford painted up in Army colors. But nothing he said made any sense at all."

Corbett said he noticed the abnormally unworn tires that were on the suspects vehicle.

"Those tires were brand new!" he said in an interview with this reporter following the mishap. "And he didn't even have a gas ration card! Why, everything about him looked fake! So, I took him in."

Trooper Corbett, who suffered minor injuries while defending the courthouse, remains in stable condition at the Uniontown Hospital.

Although details of the attack are fuzzy, it started in the late afternoon as Magistrate Cockthorn was interviewing the suspect. It seems an unknown number of enemy agents, in a brazen attempt to free the suspect, surrounded the courthouse.

"It was a well coordinated attack," said Fayette County Sheriff Eustis Poteet, who was guarding the courthouse at the time. "To create a diversion, they wrecked an oil tanker right in front of the courthouse. Then, they let loose a passel of pigs inside of the drug store. Those Nazis knew what they were doing!"

Adam's Drug and Assorted Sundries, 1210 North Court Street in Uniontown, is conveniently located in downtown Uniontown right next to the courthouse for all your prescription and over-the-counter needs. Plus, they have a full service soda fountain, where you can enjoy a cool and refreshing soda or malted while your prescription order is quickly and professionally filled.

Sheriff Poteet pursued the assailants, but was unable to apprehend them, although his deputy Roscoe Poteet (no close relation) fired five shots in the pursuit.

"The Harrington & Richardson thirty-two caliber is a maligned weaponed," said Deputy Poteet. "But the top break of the pistol allows for a quick reload, which can be an advantage in a sustained encounter with a lawbreaker."

Sheriff Eustis Poteet and Deputy Roscoe Poteet are recuperating at home. Flowers and get well cards can be dropped off at the Fayette County Sheriff's Department.

glass door, the running boards stripped off as if by a can opener. The entire cab of the vehicle was compressed, vice-like, between the pillars. The bench seat was scrunched inward, until Eustis and Deputy Poteet sat, still screaming, leaning toward each other like two parking lovers.

Deputy Poteet, ever willing to do his part, fired two more warning shots from his Harrington & Richardson five shot thirty-two caliber revolver. The tellers all held up their hands. But Luther Hankus, the ancient security guard, crawled out from behind his desk and walked over to the compressed Chevy. He leveled his Granddaddy's old civil war hog-leg Remington forty-four between Deputy Poteet's bleary eyes.

"Drop the rod, sonny!" Luther snarled. "And no funny stuff."

Sighing, Eustis turned off the siren.

Calmly, Blakeley walked out the side door

Outside the bank, Luther Hankus proudly displays his Grandfather's Remington .44, following his heroic foiling of the "German terrorist bank robbers." (Photograph courtesy of The Uniontown Citizen-Advertiser. Photograph by Sidney Welt)

Bailiff O'Riley recuperates in the alley beside the courthouse following the "vicious Nazi terrorist attack." (Photograph courtesy of The Uniontown Citizen-Advertiser. Photograph by Sidney Welt)"

and around to the front of the courthouse. As inconspicuously as possible for a man covered in tobacco spittle, he took the keys out of his pants pocket, opened the door to the Army Ford and got in the driver's seat. He depressed the gas pedal and hit the starter. The flathead V-eight instantly caught.

Up ahead of him, pigs, heady with their new-found freedom, were running out of Adams Drug and Assorted Sundries, where they slipped and slid across Court Street in the finest Pennsylvania crude oil. The alarm bell clanged insistently over at the Oilmen's and Merchant's National Bank. In the distance sirens suddenly erupted from nowhere, as police and firemen from all over the city were called to respond to the scene.

Blakeley backed the Ford up, gently tapping the bumper of the big Packard parked behind him. There was a slight space between the cars in the stalled northbound traffic. Blakeley tapped the horn lightly and motioned to the driver in the car beside him. The man backed his car up enough for Blakeley to pull out of the parking space. Blakeley cut the wheel hard to the left. He pulled out across the northbound lane of Court Street and headed south, waving thanks to the helpful driver. It was an illegal U-turn, but, under the circumstances, he didn't think anyone would

Leave them with acting Sheriff and Dispatcher Eunice Poteet. (Also no close relation.)
During the attack, Magistrate Cockthorn was rendered unconscious. Magistrate Cockthorn offered no comment on the subject. Also injured in the defense of the courthouse was Bailiff Quincy O'Riley of Beesontown, Pa. However, when interviewed by this reporter, Bailiff O'Riley had no memory of the incident. "The whole thing was put up just to break that spy out," said Lester Clement, who was in the galley at the time of the incident. "You could tell he was a spy, he just looked sneaky. And the way he talked back to the Magistrate! Well, his buddies, they come to get him! Them Germans was everywhere! Only you couldn't tell which ones was Germans, they was dressed in everyday clothes just like normal folk!"

Deputy Poteet demonstrates the steely-eyed glare that backed down the Nazi terrorists. Poteet favors the H&R .32 revolver, with custom grips.

Witness reports vary as to where the first shots originated. It is known that the suspect, who gave the obviously false name of "Lt. Julian Blakeley," disappeared sometime during the ensuing melee. In his flight, he kidnapped Myrtle Canker, wife of local Barber Homer Canker, proprietor of Canker's Barber Shop. Canker's Barber Shop, located on the corner of Court Street and Eighth Avenue, is open most weekdays from 9 to 5:30 and Saturday from 9 to noon. Stop in and see Homer for all of your hair care needs, or just to say "Hello!" Razor side cuts are a Canker specialty.

"Myrtle is a brave woman," said Homer. "She fought those Nazis all the way to Pittsburgh, with no other weapon than her pocketbook! It's a miracle that she's alive to tell the tale!"

Myrtle Canker escaped the enemy agents outside of Pittsburgh. She was treated for extreme mental agitation at the Pittsburgh Psychiatric Center and released.

In addition to the attack on the Courthouse, the perpetrators attempted to rob the Oilmen's and Merchant's National Bank. They were stymied through the heroic efforts of Oilmen's and Merchant's security guard Luther Hankus, who, in conjunction with Sheriff Poteet and Deputy Poteet, was able to repel the enemy agents.

either notice or care. Then he headed down Court Street, turning onto Route Forty. Once on Forty, he breathed a huge sigh of satisfied relief, and continued on his way to Pittsburgh, judiciously obeying all of the posted speed limits. It wasn't until he was entering the Liberty Tunnel into Pittsburgh, that Myrtle woke up in the back seat, and began to viciously bang him in the back of his head with her purse.

Senator Bogmire reacts to the news of Lieutenant Blakeley's Uniontown escapade.

Some of Homer Wells' hogs admire the variety of selection in the candy section of Adam's Drug & Assorted Sundries. (Photograph courtesy of The Fayette County Historical Society)

(Excerpt from *The Flight of the Boomerang*, by Elmer C. Wackmallit, continued.)

Liberty Tunnel
Mt. Washington, Pennsylvania
August 23, 1942 5:32 p. m.

"Oh, you vile, despicable man!" cried Myrtle. She swung her heavily laden purse with the all the force of the truly righteous. "Release me this instant!" However, at that instant, Blakeley was barreling down the Liberty Tunnel, trying his best to control the Army Ford. His task was complicated by the fact that he was being pummeled with Myrtle's handbag, which seemed to be where she kept all of her bricks.

"Unhand me, you Nazi monster!"

Getting up her steam, she hauled back and walloped Blakeley across the back of his head so hard that he saw double. It was only by the keenest concentration that he kept the car in its own lane. Behind him, a big red tanker truck loomed threateningly in the mirror. The big semi's brakes screeched, and the truck bumped his bumper lightly. The irate trucker laid on his horn, while simultaneously blinking his lights.

"Oww!" cried Blakeley. "Lady, quit hitting me! You're going to kill us both!"

"Don't you dare threaten me!" screamed Myrtle, slamming him again with the handbag. "I am a member in good standing of the Ladies' Auxiliary Civil Defense Corps, and I've been trained in self defense against German vermin like you!"

In the left lane, a green and yellow city bus pulled up beside the Ford, overtaking the tanker truck. Myrtle whacked him again with her brick bag, causing his eyes to lose focus. Blakeley slowed the car, crossing slightly into the left lane lane as he did so. In the other lane, the city bus swerved and slammed on his brakes, narrowly missing the Army Ford. The driver sped by, shaking his fist and blowing his horn wildly.

THE NAZI CABAL'S ATTACK ON MYRTLE'S VIRTUE

Myrtle whacked Blakeley again, just for emphasis. Blakeley went cross-eyed, as the huge tanker, still looming in the rearview mirror, honked his horn and flashed his lights. Myrtle, noticing the truck behind her, turned around in the seat. With her head and arms filling the rear window of the Ford, Myrtle began to shout at the truck driver.

"Help! Rape!" she screamed at the top of her lungs, which were considerable. "I'm being kidnapped! They're going to rape me! Against my will!"

The brief respite from attack allowed Blakeley's eyes to uncross. He shook his head to clear it, and reached under the front seat, removing a knapsack. Across the top of the bag in stenciled letters, was "U.S.A.A.C. Motor Pool Emergency Kit." Below that in smaller letters, it said, "Do not remove from vehicle." He ripped open the snaps, and dumped the contents on the front seat beside him. In the pile was a first aid kit, an ancient Webley Flare Pistol left

Eugene C. "Spike" Francis, the Ajax Tanker Lines driver. Although used in the story on the German saboteurs' attack on the Liberty Tunnel, this photograph was actually taken on the occasion of Mr. Francis' arrest by Pennsylvania State Police for smoking in a vehicle carrying flammable materials. It had previously appeared in the February 16, 1942 issue of the Pittsburgh Police Recorder.

over from World War I, a box of three flare shells, a carton of C-rations, and an olive drab pre-war style helmet. Blakeley quickly donned the helmet, pulling the strap down from the front brim and snugging it securely under his chin.

"Ohhh!" yelled Myrtle, "The nerve of that hooligan!" With great effort, she climbed down out of the rear window, huffing. "That nasty truck driver made a dirty sign at me with his finger!"

Blakeley instinctively hunched forward. Myrtle, remembering herself, took a belated swing at him, but it only glanced off the back of his helmet. Blakeley, his head clearing slightly, stepped on the gas, causing Myrtle to lurch back in the rear seat. She struggled upright. Rolling down the side window, she leaned out and gesticulated wildly at the trucker with both hands.

"There, you big oaf!" she screamed, holding up the middle finger of both hands. "There's double your own medicine!"

The trucker behind the Ford was blowing his horn steadily now. The big truck stayed within inches of the Ford's rear bumper, his lights flashing, the driver gesturing wildly at Myrtle, who was still hanging out the window screaming at him. Blakeley was impressed, in awe, frankly, of the driver's ability to so adroitly keep his huge truck under control while operating the horn and headlights simultaneously, all the while rudely signaling Myrtle. While Myrtle and the truck driver were engaged in their sign-language conversation, Blakeley took the opportunity to load the flare gun. With one hand on the steering wheel, he broke down the flare gun with the other. Tearing open the box of flare shells, he grabbed one, inserted it into the chamber, and snapped the gun shut. He was not at all sure how he would use it. Still, he took some comfort in knowing that he had the pistol, loaded, under the front seat. Under these circumstances, anything seemed possible. The enraged truck driver slammed the back of the Army Ford, harder this time. The car swerved wildly into the left lane, but luckily there were no cars there at the moment. Myrtle, balanced on the window sill of the door, screamed as she tumbled outward, her legs flailing upward and slamming into the roof of the car. Luckily, her huge girth came to her rescue, wedging her into the window frame.

Blakeley, steering wildly with his right hand, reached back with his left hand. Grabbing Myrtle's legs, he tried to pull her back in the window. In the other lane, a tractor trailer blew its air horn as it passed, missing Myrtle's head by inches. Myrtle kicked wildly, freeing herself from Blakeley's grip. Leveraging against the roof, she pulled herself upright, and worked her way back through the window.

"How dare you touch my person?" Myrtle shrieked. She climbed up on the back seat on her knees, and, leaning forward, began to methodically beat him with her handbag once again, punctuating her speech with each blow. "How . . . . dare . . . . you . . . . touch . . . . me . . . . you . . . . filthy . . . . little . . . . Nazi . . . . pervert!"

The protective helmet greatly reduced the effect of Myrtle's blows. Blakeley's eyes crossed only slightly, and his vision was just barely blurred this time.

"Lady, please!" said Blakeley. "Would you stop hitting me, for pete's sake? I'm not a Nazi spy!"

"So!" said Myrtle, cocking her purse. "You admit you're a pervert?"

Blakeley involuntarily cringed."Look, I'm Lieutenant Julian S. Blakeley of the U. S. Army Air Corps."

"That's just what a Nazi spy would say! Where's your uniform, Lieutenant Blakeley?" she asked, whapping him in the head for emphasis. "Or, should I say, Herr Blakeley?" She whapped him again. "Or is it, Von Blakenstrudel?" She hit him again.

"Oww!" moaned Blakeley, "Do you have to use your handbag for punctuation?"

"Where's the real Lieutenant Blakeley, if there is one?" She whapped him again. "Lying in a ditch somewhere?"

"I AM the real Lieutenant Blakeley!" he said. "I'm on a mission for the Office of Procurement! And this is my uniform!"

He looked down at the photographer's snappy plaid jacket. "Well, except for the jacket. I had to . . . borrow the jacket."

"You're on a mission all right! To Council Bluffs, Iowa!" She said, hitting him again, only not so hard, as if she was only hitting him out of habit.

"Yeah!" said Blakeley, momentarily taken aback. "How did you know that?" Because that's where the German terrorists are!" she screamed, and hit him so hard that his teeth ached.

"Help!" Myrtle screamed out the window at a passing Packard. "I'm being held hostage by a gang of Nazis! They're threatening to rape me this very

Grim aftermath: Demolished vehicles litter Washington Road, as their hapless occupants wander dazedly, left to fend for themselves in the wake of senseless destruction. (Photograph from the morgue files of the Pittsburgh Police Recorder, August, 1942)

POLICE PITTSBURGH RECORDER

NAZI AGENTS HIT CITY!
SABOTEURS TARGET LIBERTY TUNNEL!

DISASTER NARROWLY AVOIDED

HERO THWARTS ATTACK! *FULL STORY ON PAGE 3*

WOMAN STAVES OFF ENEMY KILLERS!

Above: Both "Spike" Francis and Myrtle Canker figure prominently in the Police Recorder story on the Liberty Tunnel incident. (Artifact courtesy of the Whitley Speale Collection of the Bone Lake Research Museum)

ANOTHER PAGE OUT OF HISTORY:
PRESS REACTION TO THE LIBERTY TUNNEL INCIDENT

(Editor's note: The following story appeared in the Pittsburgh Police Recorder, August 24, 1942, morning edition.)

LIBERTY TUNNEL ESCAPES CERTAIN DESTRUCTION!

Tragedy was narrowly averted yesterday, when a plot by German sympathizers to blow up the Liberty Tunnel was thwarted, due to the vigilant diligence of the alert citizenry of Mount Washington, and the heroic efforts of one Mrs. Myrtle Canker.

minute!"

The trucker, still following, saw Myrtle hanging out the window, and sped up. Blakeley saw him in the mirror, and sped up also, so that the trucker was only able to tap the rear bumper of the Army Ford lightly, causing only a slight waver in its course. Myrtle gestured profanely at the approaching truck, then quickly ducked back into the rear seat of the car.

"Are you and that driver in cahoots?" she asked, hitting him square on the top of the helmet again. "Are you in the same spy ring?"

"Lady, I didn't know anything about that truck driver until you decided to give him the finger!" Blakeley said.

"Don't you dare use that kind of language with me!" She repositioned herself on the back seat, and Blakeley hunkered down, awaiting the incoming barrage. "Don't . . . . you . . . . dare . . . . use . . . . that . . . . language . . . . with . . . . me!"

"Lady, I'm warning you, I can't drive if you keep hitting me like that!"

"Then stop the car, and let me out, you, you . . . . kidnapper!"

Eugene C. "Spike" Francis' official "Ajax Tanker Lines" hat badge. Artifact photograph courtesy of Mr. Francis' grandson, Mr. Eugene C. "Spike" Francis III.

"Nothing would make me happier, but if you hadn't noticed, we're in the middle of Liberty Tunnel, and there's a twenty-ton tractor-trailer trying to climb up my tailpipe!"

"You can play innocent if you like," Myrtle said petulantly. "By the way, your accent is very good. It doesn't sound at all like a Nazi. They must have a very good spy school over there in Hitlerville."

"Like I told you before, I'm not a spy. Well, I guess I sort of am, but-"

"I knew it!" she shrieked and swatted him across the back of his head again.

"But I work for our side! Now, quit it, will you?"

"If you're on our side, why did you try to murder that poor policeman back in Fairchance?"

"I didn't try to murder him You just heard one side of the story, and you haven't even got that right!"

"Well, duh, stupid little me! Let's see, who am I going to believe, a Pennsylvania State Trooper or a German Nazi pervert spy?"

The truck pulled up on Blakeley's bumper again, blowing his air horn and flashing his headlights.

Mrs. Canker, homemaker and wife of Homer Canker, proprietor of Canker's Barbershop in Uniontown, was taken hostage early yesterday afternoon by the Nazi saboteurs in their sweep of terror through southwestern Pennsylvania. After attacking the courthouse and attempting to rob a bank in Uniontown, the pro-German thugs proceeded westward on Highway 40 to the Liberty Tunnel, where they tried to hijack a gasoline tanker truck, owned by the Ajax Tanker Lines of Du Bois, Pennsylvania. The vicious schweinhundes planned to abandon the tanker in the tunnel and set it afire with a flare gun, resulting in a tragic loss of life and a vital blow to the transportation system of Pittsburgh.

However, Ajax driver Eugene C. "Spike" Francis had other ideas. He was able to resist the wily saboteurs, and drove the burning rig out of the tunnel at great personal risk, reaching the parking lot of Ajax Riverboats of Mount Washington, Pa. (not affiliated with Ajax Tanker Lines). After repeated attempts by Mr. Francis to extinguish the fire, the truck exploded. Incredibly, there were no injuries, due in large part to the diligence of Mr. Francis in driving the burning truck to a relatively safe and remote location. However, the truck, and several automobiles belonging to employees of Ajax Riverboats, were declared a total loss.

"If I ever see that little sawed-off runt again, I'll tear him to little bitty pieces!" said Mr. Francis, in an apparent reference to the ringleader of the Nazis, identified by Mrs. Canker as a Colonel Blakenstrudel of the elite German "S.S." or Storm Troopers. "And the same goes for that nutty wife of his!"

Authorities are uncertain as to the

total number of saboteurs, although from the statements of Mr. Francis, they apparently included at least one woman. Additionally, Mrs. Canker has indicated that Col. Blakenstrudel had at least one accomplice who assisted in the hijacking of the Ajax tanker truck.

"He was in cahoots with that big Heinie ape! "said Mrs. Canker. "They thought they were pretty smart, but they didn't fool me! They wanted to blow up the tunnel, and me with it! After they'd had their way with me, of course."

From information gained by the Pennsylvania State Police, the saboteurs are part of a ring of German and Japanese sympathizers, involved in both espionage and sabotage activities, centered in the Council Bluffs, Iowa area. Mrs. Canker, who was brutalized for hours by the vicious and bloodthirsty traitors, said she was lucky to flee from Blakenstrudel. Mrs. Canker stated that she was in constant fear for her life until she overcame him and escaped. Blakenstrudel, his plans thwarted, then made his getaway. The whereabouts of the Colonel and the members of his sabotage ring are currently unknown.

"He was an animal," sobbed the distraught Mrs. Canker. "I barely escaped with my virtue intact. He wanted me, and he wanted me bad. You could see it in his eyes."

"If these dirty goose steppers had been successful, they would have closed down the Liberty Tunnel for months. Think

The end of the tunnel was in sight, a small bright speck on the otherwise dark horizon. Blakeley, who was already going dangerously fast in the narrow confines of the tunnel, gave the Ford a little more gas, inching away from the big tractor-trailer.

"Honest, just as soon as we get out of this tunnel, I'm pulling over and letting you out, okay?"

"Why should I trust you?"

"By the way, how did you get in my car anyway?"

"Well, I guess I'm here because you kidnapped me! I mean, don't you know that? I thought spies were supposed to be smart?"

"Look, lady, I don't even know who you are, much less how you got in this car, but-"

"I am not in the habit of giving my name to enemy agents."

"Well, believe me when I say that I want you out of this car in the worst way."

"If that's the case, Mister Smarty Pants Von Blakenburger, then why did you kidnap me?"

Blakeley, not replying, continued to drive

Mr. Wallace P. Ruttles of the Mount Washington Civil Defense, who coordinated the response to the German Fifth Column's attack on the Liberty Tunnel. (Pittsburgh Police Recorder)

doggedly. Myrtle was quiet, petulant. "You mean, you don't want to rape me?"

"It's nothing personal. But no. I don't want to rape you."

"You're just going to let me go?"

"Just as soon as we get out of this tunnel. Just don't hit me with that purse anymore. I can't take it."

"Okay," she said guardedly, "but don't expect any quarter once you let me out of this car. It's my patriotic duty to stop your evil plan if I can."

At the end of the tunnel, they emerged into the bright light of day. They came to a stoplight, which was thankfully green. Blakeley went through the light just as it changed to yellow. Turning right, he drove onto a narrow little single lane brick street that ran along the river front. There was a street sign which read, "Mount Washington Road."

SPIKE PUTS IN HIS TWO CENT'S WORTH

Blakeley looked in the rearview mirror. The tractor trailer had turned the corner, running the red light. This action forced a battered taxi into the opposite lane, where it collided with a Dodge delivery wagon. Black smoke rolled out of the stack of the big truck as the driver opened the throttle. The right rear wheel of the trailer thumped over the curb, narrowly missing a group of men in hardhats standing at the corner, waiting to cross. The men jumped back, cursing and waving their fists at the driver. Blakeley drove on down the narrow brick street.

"I thought you were going to let me out?" said Myrtle.

"I've got to find a place to pull over," said Blakeley.

"Stop!" yelled Myrtle. "Right here! What's wrong with stopping right here in the middle of the street?"

"That trucker's still behind us," said Blakeley. "If I stop in the middle of the road, he'll run right over us!."

Blakeley drove down the road, until he spied a muddy pull off ahead, next to a dock yard. There were numerous cars parked in the area.

"That looks like a parking lot for the tugboat workers," said Blakeley. "I'll pull in there."

Blakeley drove into the muddy parking lot. He opened the door, got out of the car, and opened the back door. Myrtle stared at him. Her lower lip began

what that would have done to the war effort," said Mount Washington, Pa. Civil Defense Volunteer Coordinator Wallace P. Ruttles, who responded to the scene of the debacle, along with units of the Mount Washington Police Department, and the Mount Washington Volunteer Fire Department. Mr. Ruttles said that Mrs. Canker and Mr. Francis are to be honored in a private ceremony today by the Mayor of Mount Washington for their bravery in preventing the Nazi's destruction of the Liberty Tunnel. The ceremony is expected to occur shortly after Mrs. Canker's anticipated release from the Pittsburgh Psychiatric Center later this afternoon.

Although Mayor O'Rourke was unavailable for comment, his office issued the following statement: "Let this action by these brave citizens stand in mute testimony to those denizens of democracy who would threaten our way of life. Cower in the corner like the verminous traitors you are, until the bright light of freedom reveals your treacherous den! But never, never, dare to work your vile misdeeds in Mount Washington again, for her citizens will rise up once again to smite you!"

Excerpt from: The Pittsburgh Police Recorder, August 24, 1942; Morning Edition

In this photograph featured on the front page of the Pittsburgh Police Recorder, Myrtle Canker is supposedly shown demonstrating her pluck for some adoring fans.

The cutline reads as follows: "Mrs. Myrtle Canker, the unlikely heroine of yesterday's Nazi terrorist attack, demonstrates her defensive handbag technique before a spellbound crowd of dockworkers and other onlookers at the site of the Axis hooligan's attack in the parking lot of Ajax Riverboats in Mount Washington, Pa., while one of the destroyed vehicles smolders in the background."

The photograph, purportedly taken by Pittsburgh Police Recorder Staff Photographer Ralph A. Pauling, is almost certainly a fake. The lady in the photograph is not Myrtle Canker, who, as noted previously, was a blonde. The lady is most likely Senora Emanuella Luciardo, an exotic dancer who performed her famous "Dance of the Seven Aprons," under the stage name of "La Chicarina," on a regular basis for the disinterested patrons of Enrico's House of Spaghetti in Mount Washington, Pa.

to quiver.

"Okay," he said. "There you go! You're free! Now, get out of my car!"

Myrtle sat in the back seat, looking at Blakeley. She looked like she was going to cry.

"Are you just going to leave me here?" she asked plaintively. "Out in the middle of nowhere?"

"Cripes, lady, isn't that what you've been screaming about for the last twenty minutes? 'Let me go! I'm being kidnapped by Nazi perverts!' Well, I'm letting you go! Now, go!"

She looked around the shipyard parking lot. "Who knows what kind of ruffians hang out in this parking lot? This is the waterfront, after all. Aren't you the least bit concerned about my safety? There could be-"

"What?" screamed Blakeley. "There could be what? Rapists? Nazi spies? Perverts? Don't worry. You'll still have your pocketbook to defend yourself with! Now, for the last time, lady, get out of the car!"

Blakeley reached into the car and grabbed Myrtle by the wrist. He pulled, but it was no use.

Pittsburgh Police Recorder photographer Ralph A. Pauling, with the tool of his trade. Pauling, unscrupulous and devious by all accounts, shamelessly duped his gullible readers into believing that they were viewing photos of the genuine Myrtle Canker, instead of the notorious "La Chicarina." ((Photograph courtesy of the Whitley Speale Collection of the Bone Lake Research Museum)

There was too much of her and too little of him. Then there was a screech of brakes, and the roar of the Peterbilt's Cummins diesel being seriously downshifted. The big red tanker pulled into the parking lot, splashing through the mud puddles. Steam hissed up from under the truck as it skidded to a halt in the gravel, just inches behind the Army Ford. The door opened, and the driver jumped out, his size fourteen cobbled work boots landing with a splashy thud in the muddy gravel. He reminded Blakeley of Frankenstein's monster, only with a pot belly and slightly better surgery. He wore a dirty "Ajax Tankers" coverall that looked like it hadn't been washed since the stock market crash. His sleeves were rolled up over tattooed biceps as big around as Myrtle's thighs. He grabbed a grimy tattered company hat from his crew cut head and threw it on the ground. His Neanderthal brow creased even lower, if that was possible, as he stomped toward Blakeley and Myrtle, pulling a cigar the size of the Hindenburg out of his scowling mouth. He rubbed out the cigar on the fender of his truck, and dropped it in the dirt. The oily name tag on his grimy coveralls said "Spike."

Spike marched up to Blakeley like the Blitzkrieg marching into Poland. He poked a giant hairy grease-caked finger in Blakeley's chest.

"What kind of pissant are you, buddy, that you can't control your wife there?" asked Spike, pointing at Myrtle in the back seat. He poked Blakeley in the chest two more times, then again, to emphasize his point.

"She . . . she's not my wife," said Blakeley.

"I don't care if she's your old grandmammy!" said Spike, pushing Blakeley in the chest three more times. "I ain't interested in your personal life. You know what she was doin' in the back window of your car while you were lollygaggin' through the tunnel? Making dirty signs at me with her fingers, that's what!"

"Look," said Blakeley. "I don't know this woman. I never saw her before. Then, not fifteen minutes ago, she jumps up out of the back seat of my car and starts beating me over the head with a knapsack."

"It was a handbag," yelled Myrtle from the back seat.

"You keep out of this, lady!" said Spike.

"It felt like a knapsack," said Blakeley. "Full of rocks."

"What are you, sonny, some kind of wise guy?" asked Spike, poking Blakeley in the chest three more times. "And what's up with this soldier hat you got on, at least it sort of looks like a soldier hat. You playing soldier?"

"I'm not playing!" said Blakeley, straightening up to his full five foot six inches, and looking Spike straight in the chest. "I'm on official business for the U. S. Army Air Corps, and you are interfering with an official Air Corps mission!"

Spike poked Blakeley in the chest three more times. "If you're a soldier, where's your uniform?"

"That's what I asked him!" said Myrtle.

"You stay out of this!" Blakeley and Spike yelled in unison.

"If this is for my benefit, you can both stop, because you're not fooling me one bit!" said Myrtle. "I know you're in cahoots!"

"For that matter, if you're in the Air Corps, why aren't you in an airplane?" asked Spike. "Whoever heard of a fly-boy in a Ford?"

"Look, it's been nice chatting with you, but I've got a mission–"

Spike poked him in the chest three more times. "I saw John Wayne in 'Flying Tigers!' He wasn't riding around in a Ford! He had a P-40! And he wasn't driving like a crazy maniac through the Liberty Tunnel, letting his wife give the dirty finger to an innocent truck driver who's hauling motor fuels that are critical to the war effort! No! He was shooting down Japs! So, you tell me, you little sawed-off zoot-suit pipsqueak, why aren't you over in Japland, shooting down Japs?"

This is the photograph from which Ralph A. Pauling purloined the image of "La Chicarina" for his notoriously misleading front page photo of "Myrtle Canker." This photo caption from a story of the previous May reads: "Local dancer arrested. Señora Emanuella Luciardo was arrested by officers of the Mount Washington Municipal Police Department today for performing her notorious "Dance of the Seven Aprons" before a group that included minor children." The Police Recorder dutifully blacked out the patron's faces to "protect the innocent." (Pittsburgh Police Recorder, May 8, 1942)

"He doesn't have time for that!" yelled Myrtle from the backseat of the Ford. "He has to blow up Council Bluffs!"

Spike shook his head sadly.

"I thought I told you to shut her up!" he told Blakeley, and then he hit him on the jaw with such force that Blakeley was jerked off his feet like an umbrella in a tornado. He flew six feet through the air and landed with a splash in a huge grey mud puddle. Spike advanced on Blakeley, who was lying on his back in the puddle. Blakeley sat up and shook his head groggily. Spike leaned over Blakeley, grabbing him by the collar and lifting him up to hit him again.

8. MYRTLE'S REVENGE

The grim result of the Nazi Saboteurs' attempted strike on the Liberty Tunnel is evident in this photgraph taken on the evening of August 23, 1942, from a vantage point just southwest of Washington, Pa. (Photograph courtesy of the Whitley Speale Collection of the Bone Lake Research Museum)

(Excerpt from *The Flight of the Boomerang*, by Elmer C. Wackmallit, continued.)

"Hold it right there, both of you!" yelled Myrtle. She was stepping out of the back seat, smoothing down her ruffled dress with her left hand. In her right hand was the flare gun, pointed at Spike's chest.

"You two think you're so smart!" snarled Myrtle, waving the flare gun at Spike as she advanced toward him. "You think that just because I'm a woman, I haven't figured out your little scheme?"

Spike dropped Blakeley back in the puddle, and held up his hands as he turned to face Myrtle.

"Now, look, lady," he said softly. "You want to put that thing down. I didn't mean to hit your husband. It was just a little misunderstanding, right buddy?" He looked down at Blakeley, who stood up unsteadily beside Spike. Spike wiped ineffectually at the water on Blakeley's coat.

"You don't want to do something you'll be sorry about later," said Blakeley.

"You didn't think I saw you put this peashooter under your seat, did you?" she asked Blakeley, smiling gleefully. "Then your buddy here conveniently shows up with a truck full of gasoline. Why, together, you two could have blown up the Liberty Tunnel, then drove on down the road in your little fake Army Ford, and nobody would have ever known any better!"

"Be very careful with that, ma'am," said Blakeley. "It's loaded."

"What's that medal they give you back in Hitler-land for something like this? The Double Cross?"

"I believe you're thinking of the Iron Cross," said Blakeley.

"That's it," said Spike, glaring at Blakeley. "Egg her on!"

"Keep your hands up!" growled Myrtle. "Well, I got news for you. There's not going to be any kind of cross for Herr Blakerhoofer, or his buddy Herr whatever your name is."

"Spike," said Spike.

"What did you say?" asked Myrtle, waving the

gun menacingly in Spike's face.

"Spike," said Spike. "It's my name. Spike. Ma'am."

"I don't care what your name is!" hissed Myrtle. "What makes you think I give a fart in a windstorm what your name is?"

"That's just the kind of language I would expect from somebody who flips me the dirty finger sign from the back seat of a Ford," said Spike, his feelings hurt.

Myrtle drew in her breath, puffing up with anger. "Look here, buster, you've got some nerve-"

"Listen, lady," said Blakeley, interrupting her. "There's only one shot in that gun. Why don't you just hand it to me real slowly?"

Blakeley took a slow step toward Myrtle, holding out his hand tentatively.

Myrtle brought the flare gun to bear on Blakeley. "You make just one more move toward me, you pantywaist Heinie slime ball, and I'll blow a hole in you so big your Nazi buddy there can drive his trailer truck through it!"

"Hey, what's got into you?" yelled Spike. "I ain't no Nazi!"

"The same goes for you, you goose-stepper!" Myrtle swung the flare gun back to Spike.

"You've sure got a nasty mouth on you, lady," said Spike.

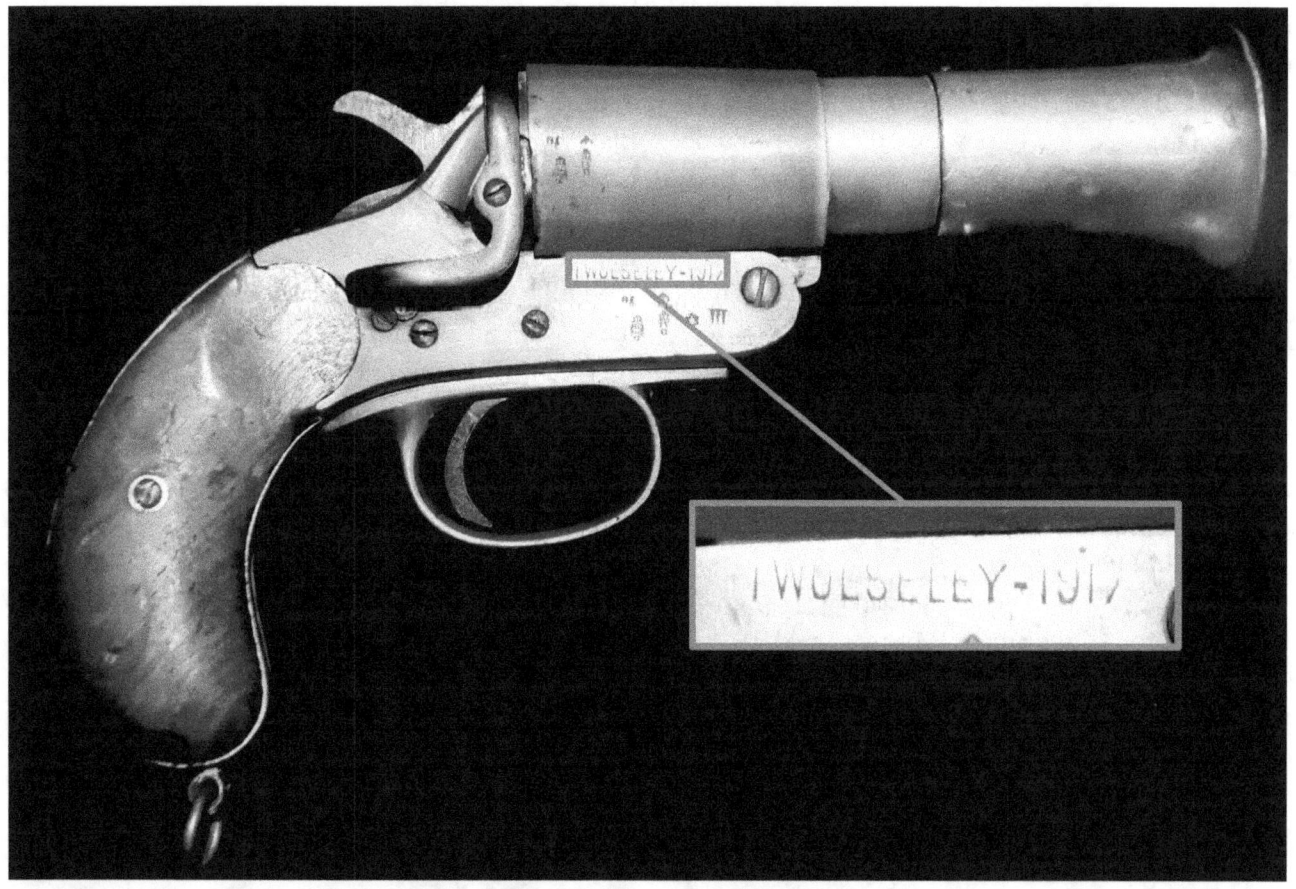

This World War I Flare Pistol, currently on display in the Whitley Speale Collection of the Bone Lake Research Museum, is purported to be the pistol used by Myrtle Canker to thwart the German Saboteurs in the famed "Liberty Tunnel Attack." This supposed "fact" is vehemently disputed by none other than Halloway Bumpsteed, Jr., in his now legendary work *They Might Have Been Giants: Misadventures, Blunders, & Colossal Failures in Aviation*. Bumpsteed rightly notes that the pistol in question, while indeed a Webley Signal Pistol, was manufactured under license by the Wolseley Sheep Shearing Company, whose mark can be plainly seen on the frame of the weapon. Bumpsteed notes that the Wackmallit text specifically states a "Webley Flare Pistol" was used in the attack. Bumpsteed has subsequently sued the Whitley Speale Collection, demanding removal of the flare pistol as fraudulent. Whitley Speale has countersued Bumpsteed for slander and defamation of character. At the time of publication, both lawsuits remain unresolved. (Photograph courtesy of Picasa, with kind permission of the Whitley Speale Collection of the Bone Lake Research Museum.)

"I don't waste my good manners on Nazi terrorists!" said Myrtle.

"Like I said, you've only got one shot," Blakeley said calmly. "You can get one of us, but you can't get us both. Why don't you put the gun down?"

Myrtle waved the gun back and forth, first at Blakeley, then at Spike.

"Of course, if you did decide to shoot somebody, this looney-tune here would be a good choice," said Spike, nodding his head at Blakeley.

"Oh, yeah, you think so?" Myrtle swung the gun back to Spike, holding it with both hands now. "If I did shoot him, what do you think you would do to me?"

"Lady, that's not exactly fair." Spike smiled sheepishly. "I mean, I don't think there's a right answer to that question."

"Oh, no?" Myrtle eyed Spike menacingly. "I know what you'd do. I know exactly what you'd do! You wouldn't do a thing in the world to me, that's what you'd do. You know why? Because you'd be too busy, that's why."

Spike looked at Myrtle uncertainly, then at Blakeley.

"You want to help me out here?" he asked. Blakeley shrugged his shoulders.

"You want to know why you'd be too busy?" asked Myrtle sweetly.

"Okay," sighed Spike. "I'll bite. Why would I be too busy?"

"Because, you stupid Hitler-loving moron," screamed Myrtle, swinging the flare gun around and pointing it at his truck. "Your truck's on fire!"

"No, lady! Don't do that!" Spike screamed as Myrtle pulled the trigger. The flare gun cracked, sending a spiraling plume of yellowish smoke in a graceful arc toward the parked fuel truck. The flare bullet crashed through the driver's front window of the bright red Peterbilt tractor. There was a muffled "Whoof!" and a brilliant light from within the cab, followed by billowing yellow smoke which rolled out of the broken cab

window. Spike ran to the truck, as flames of bright orange began to roll out of the side windows.

Myrtle laughed maniacally, watching in delight as Spike ran around the burning cab to a side panel on the trailer. He got out a fire extinguisher and pulled the pin, aimed it into the cab, and pulled the trigger. A white plume of extinguishing agent poured onto the brilliant flames, with little effect.

MYRTLE DISARMED

Blakeley stood, dumfounded, looking at Myrtle. She suddenly looked over at him, as if she finally remembered that he was there. She grimaced, and pulled the flare gun around to bear on him.

She pulled the trigger once, and the flare gun clicked. She pulled the trigger once again, and then a third time. Nothing happened. She looked at Blakeley, then at the flare gun, a look of growing horror on her face. She screamed, dropped the flare gun, and ran out into the street, dodging across the traffic.

Dully, Blakeley registered the screech of brakes, the crash, and the sound of shattering glass as a car swerved into the opposite lane to avoid hitting Myrtle.

Blakeley wiped at the muddy water on his sleeves, but the jacket was soaked. He took the jacket off and laid it on

The day after the Liberty Tunnel altercation, the smoking remains of the Ajax tanker still smolder, guarded by a lone Pittsburgh motorcycle policeman.

TWO FIGHTER CONCEPTS THAT ULTIMATELY SUCCEEDED: THE NORTH AMERICAN NA-73X AND THE SEVERSKY XP-41

The plethora of attempted fighter and pursuit designs illustrated in the previous chapters (with the sole exception of the Lockheed P-38 "Lightning") would seem to indicate that nobody in the U.S. aviation industry during World War II could get it right. Nothing could be further from the truth.

the back seat, closing the rear door. Across the road, frantic dock workers were running out of the line shack, heading toward the parking lot. Crewmen on the moored barges pointed frantically in the direction of the burning truck.

More traffic screeched to a halt out in the road. In the distance, he could hear Myrtle screaming at

Above: A British North American NA-73 "Mustang Mk I" being test flown in U.S. markings at North American's plant in October, 1942. Inset: The original NA-73X. Even with the cowling lines of the Allison V-1710 engine, the aircraft distinctly foreshadows the P-51. (Photographs courtesy of the Library of Congress Office of War Information Collection.)

THE NA-73X

The NA-73X was developed in response to a request by the British government in 1940 for a fighter bomber to augment their Spitfires and Hurricanes. Initially requested to produce Curtiss P-40s for

the top of her lungs at the stopped traffic. Something about raping Nazis and saboteurs trying to blow up the Liberty Tunnel. In the distance, across the river, he could hear the sirens. There were quite a few, police cars, and fire trucks, he noticed, but there was also a big siren winding up slowly that Blakeley thought must be an air raid siren.

"My Girl," a North American P-51, just before flight release from Iwo Jima. The P-51 served with distinction in both major theaters of the war. (U.S. Air Force photo)

Blakeley picked up the Webley Flare pistol, opened the chamber, and ejected the used cartridge onto the ground. He got in the Army Ford and closed the door gently. He unhooked his chin strap, took off his helmet, and packed the helmet, flare gun, the carton of C-rations, and remaining shells back in the knapsack with the stenciling across the top of the bag which said, "U.S.A.A.C. Motor Pool Emergency Kit," and below that, in smaller letters, "Do Not Remove From Vehicle." He placed the knapsack back under the front seat.

Out in the road, traffic was backing up, as Myrtle had fainted. A good samaritan motorist was attempting to revive her with mouth to mouth resuscitation. Blakeley carefully tucked his black tie between the second and third buttons of his khaki shirt. Spike, his fire extinguisher emptied, cursed, kicking the door of the GMC with his hobnailed boot. He threw down the fire extinguisher, picked up his company hat, dusted it off on his knee, and put it back on his head. Out in the street, Myrtle finally came to. She beat her rescuer over the head

Britain under license, North American rightly maintained that they could produce a better fighter than the P-40. The NA-73X utilized the same Allison V-1710 engine as used in the Curtiss P-40, but the NA-73X incorporated many improvements, the most notable being a high efficiency laminar flow wing. Because of the poor high altitude performance of the Allison engine, the British used the "Mustang Mk I," the production name for the NA-73X in low level ground attack duties. Ultimately, the aircraft was purchased by the USAAC and designated the P-51 "Mustang." With the incorporation of the Packard V-1630 engine in the P-51-B series, the high altitude limitations of the Allison engine were overcome, and the P-51 achieved great success in every theater of World War II. It is considered by many to be the finest piston fighter ever built. With the addition of long-range drop tanks, the P-51s escorted U.S. bombers on air raids deep into Germany, achieving outstanding success in defending the bombers from enemy fighter attack.

THE SEVERSKY XP-41

Although only a single prototype of the Seversky XP-41 was built before cancellation of the project, the XP-41 is significant primarily in the choice of powerplant for the aircraft, and the subsequent influence on later designs of the company. Powered by a Pratt & Whitney R-1830 radial engine producing 1,200 horsepower. The engine featured a

Above and right: Two views of the Seversky XP-41. Only one prototype was built before the project was cancelled. (US Air Force photos)

with her handbag full of bricks, screaming that she was being raped by Nazis. Spike noticed his cigar lying in the dirt. He picked up the cigar. Holding it up to the flames coming out of the driver's side window, he lit it up. Blakeley started the Army Ford. Spike, puffing on the cigar, looked up as he heard Blakeley's car starting.

two-stage supercharger developed by Boeing Aircraft Company initially for the B-17 Flying Fortress, which gave the XP-41 a significantly improved high altitude performance, a feature which was lacking on most contemporary U.S. designs (notably, the NA-73X above).

Concurrently with the XP-41, which was actually developed from the last Seversky P-35 pursuit, Seversky was developing an all new fighter, the P-43 "Lancer." Now known as Republic Aircraft, the company was requested to proceed with the development of the P-43 as the superior model.

Blakeley pushed in the clutch, put the gear shift lever into first, and pulled out into the road.

"Hey, you!" yelled Spike. "Come back here!"

Then Spike took off running down the muddy parking lot in pursuit of the Army Ford.

In the middle of Mt. Washington Road, Myrtle heard Spike yell. She was momentarily distracted from

Above, and left: The Republic P-43 was one of the few high altitude fighters that the USAAC possessed at the outbreak of World War II, but it was severely handicapped by the lack of self-sealing fuel tanks and inadequate armament. (US Air Force photos)

Although the P-43 never lived up to its expectations, it high altitude performance was considered noteworthy, and it served for a time with Claire Chennault's "Flying Tigers" in China before being remandered to the Chinese Air Force.

flailing her unfortunate rescuer. She looked up in time to see Blakeley pulling out of the entrance to the parking lot, Spike running behind him with huge flying leaps.

"Hey!" yelled Myrtle to Blakeley, "You can't just leave me here!"

The Republic P-47 remains one of the most significant fighter aircraft of all time, and one of the most massively produced, with over 15,000 examples. Few remain today, as the majority were scrapped at the end of World War II. (US Air Force photo)

Myrtle gave the rescuer a parting whack on the head, and took off down the highway. Blakeley looked both ways for traffic, then calmly took a left turn onto Mt. Washington Road. He accelerated smoothly across the Mt. Washington Bridge.

Spike ran after him, but he was no match for the smooth flat-head V-8 of the Army Ford. Spike stopped in the middle of the bridge. He bent over, hands on his knees, breathing heavily. Myrtle came up behind him, yelling at Blakeley. The Army Ford receded in the distance, a light puff of white smoke coming out of the tailpipe as Blakeley changed from first to second gear. She stopped abruptly, as if noticing Spike standing on the bridge for the first time.

"Oh, no," she said quietly, almost under her breath. She stopped, turned and walked back up the road, looking

By the time the type reached active service with the USAAC, it was already considered obsolete. Republic Aviation was tasked with developing a totally new aircraft, utilizing the Prat & Whitney R-2800 radial engine, producing 2,000 horsepower. The result was the Republic P-47 "Thunderbolt" series. Although capable of being outturned by both the German BF-109 and FW-190 at lower altitudes, the P-47 was incredibly tough and well armoured, and the Pratt & Whitney R-2800 engine was extremely reliable under wartime combat conditions. Tactics were soon developed to utilize the strengths of the P-47 in combat, most notably its climb, dive, an high altitude performance.

By the end of World War II, 15,686 P-47s had been produced, and 15,875 P-51 Mustangs. Both aircraft remain among the most highly regarded piston engined fighters of all time.

Above, right: A P-47 equipped with a long-range belly tank makes a low pass over an English Aerodrome. The "drop tanks" allowed both the P-47 and the P-51 to escort long range bombers deep within enemy territory. The tanks were then dropped before engaging the enemy in combat, hence the term "drop tank."

over her shoulder at Spike. Spike looked up, saw her, and began to run after her.

"Help!" screamed Myrtle, as she ran back up the

The Quartermaster's Store on State Street, Pittsburgh, where Blakeley was able to refurbish his disheveled uniform. (Photograph from the morgue files of the Pittsburgh Police Recorder, June, 1941)

road. "Nazi rapists!"

Blakeley stopped at the light on 10th Avenue. He heard a soft thump, as if from a far away explosion. He looked in his rear view mirror, where he could see a giant fireball rising up in the sky, turning into thick black smoke. It appeared to be about where the tanker truck was parked. The light turned green and Blakeley turned left, heading toward Pittsburgh.

Just as he passed Third Avenue, a line of police cars came roaring across the 22nd Street Bridge. Good citizen that he was, Blakeley pulled over to the side of the street to let them pass. He started back out into the street, but almost immediately a line of fire trucks came rolling off the bridge, and Blakeley pulled to the side of the road once again. After the parade was over, Blakeley crossed the 22nd Avenue Bridge and finally drove into Pittsburgh.

Remembering the State Trooper's remark about the Quartermaster's Store on State Street, Blakeley stopped at an Esso station and got directions. He was hoping that he could replace his missing jacket, and maybe even get a new pair of trousers, as his were fairly messy by now.

He found the store easily enough, but unfortunately they were out of jackets and trousers. He bought a new shirt, and, upon presenting the proper identification, he was able to purchase a shiny new set of Lieutenant's bars. As he was walking out of the Quartermaster store, he noticed a mannequin in the window that was modeling the new A-2 Flying Jacket that was just being issued to the Army Air Corps. He went back to the counter and inquired about the flying jackets. They had these in stock, in his size, and, now that he thought about it, he would much prefer the flying jacket. Upon further inquiry, he discovered that just down the street, there was a nifty little dry cleaners where you could have your pants pressed while you played putt-putt golf. Blakeley shot a 54, which was three under par. Plus, by getting a hole in one on the third hole, he got his tie cleaned for free.

Above: A few of the thousands of posters that reflected the tenor of the times, as Blakeley progressed on his pilgrimage of discovery. (Images courtesy of the National Archives and Records Administration.)

9. ONWARD TO ZANESVILLE

Entryway to Grackle's Blue Moon Motor Inn, nexus of social activity for personnel of the Zanesville School of Aviation Maintenance & Beauty technology. (Photo courtesy of Dr. E.C. Whimpington)

THE BLUE MOON

(Excerpt from *The Flight of the Boomerang*, by Elmer C. Wackmallit, continued.)

Six hours later, and barely able to keep his eyes on the highway, Blakeley rolled into Zanesville, Ohio. He checked into the Blue Moon Motor Inn, just across the street from the Zanesville Municipal Airport. He slept like a log.

The next morning, he woke up rested. He pulled on his trouser pinks and a Tee-shirt. Without bothering to put on his socks, he slipped into his regulation brown oxfords. He pulled back the cheap curtains, letting the bright light of the rising sun pierce into the gloomy room. After splashing some cold water from the basin over his face, he opened the front door of his motel room and breathed deeply of the fresh late summer air. It was only then that he noticed his car. During the night, someone had jacked up his Ford and stolen all the tires.

Blakeley went to the motel office and knocked loudly on the door, but no one answered. He sniffed the air. There was the distinct smell of bacon on a grease-caked grille. Blakeley followed his nose across the Motor Inn parking lot. On the other side, tucked into a bristle of wild shrubbery on an otherwise abandoned lot, he found the Blue Moon Diner, a dilapidated turn of the century railroad passenger car sitting lopsided on crumbling cinderblocks. The many layers of peeling paint were all that kept the crumbling wooden sideboards from falling off the sides and into the weeds. From somewhere on the roof, an exhaust fan pumped a voluminous cloud of noxious bacon and hash brown fumes into the crisp morning air.

Blakeley walked up the rickety steps. He turned the worn brass knob, pushing open the ancient door, which emitted a loud creak. Inside the diner, it was at least twenty degrees warmer than outside, and the bacon and hash brown smell was like a solid wall. The train car was narrow, and on the patron side of the establishment, there was only room for a row of stools and a narrow aisle. At the counter sat a group of teenagers, engaged in a lively babble of conversation. As Blakeley noisily closed the door, they all turned to look at him. The room fell immediately silent.

These teenagers appeared to Blakeley to be an odd group. Every boy in the diner was dressed in matching sky-blue coveralls with an emblem and writing stitched in yellow thread over the left pocket. Stitched on the backs of their coveralls in the same coarse yellow thread were the letters "Z.S. of A.M. & B.T." above a logo of a hair dryer with wings. These boys were interspersed with girls, all dressed in the latest fashions, all with elaborate hair and long ruby fingernails.

Behind the grill, a beefy guy in a greasy brown apron scraped the bacon crud off the grill. He grabbed a spatula full of lard and threw it on the grill, where it exploded with a loud series of pops and sizzles. He wore an incongruous chef's hat the same shade of brown as his apron. He turned his head, flicking his cigar ash onto the freshly greased grill. Blakeley noted that the deep brown apron and chef's hat had probably once been white.

"Take a stool, buddy!" he shouted over the popping and cracking of the frying lard. "I'll be wit' ya in a minute."

"I can't buy breakfast just yet," said Blakeley. "I left my wallet in my room."

There was a sniggering of laughter from the other end of the counter. One of the boys, a lanky kid in coveralls with a protruding Adam's apple and a face full of acne, yelled, "What's your room number?"

The whole group erupted in a gale of raucous laughter.

The Blue Moon Diner is shown in this photograph taken shortly before its demolition. The building in the background is the main building of the Blue Moon Motor Inn, which also housed Miss Kitty Kat's Tophat Club. The presence of the this building dates the picture to before its destruction in August, 1942. The gentleman in the window is widely believed to be none other than Hoppity Hooper himself. (Image courtesy of the private collection of Dr. Manfredd von Goetzzenberger of Das FleugelWerkes.)

The coverall logo that Lieutenant Blakeley encountered in the Blue Moon Diner. Oddly, although it was required for all of the aviation maintenance students of the school, none of the beauty technology students were required to wear it. (Artifact courtesy of the Whitley Speale Collection of the Bone Lake Research Museum)

The big guy behind the counter deftly picked up an egg with his spatula and hurled it all the way to the end of the diner, where it hit the wall with a loud crack. As it slowly oozed down the wall, Blakeley could see the ancient imprints of hundreds of previous egg strikes, each with its own petrified ooze trail.

"I told you hyenas to keep it down in here!" the big man yelled.

"Aww, Hoppity," brayed the lanky kid. "We were just pulling the guy's leg!"

"It's Mr. Hooper to you, Lansberg!" Hoppity pointed his spatula at Lansberg, the lanky kid, squinting his eye as if he was taking aim."And a dime for that egg is going on your tab!"

"No, you see," Blakeley tried to start over. "There's no one in the office and–"

"Hilda don't get out

of bed before two o'clock," Hoppity, still flipping his eggs, interrupted Blakeley. He didn't look up.

"She doesn't like to drink in the morning," said a girl sitting at the counter, turning to Blakeley and smiling brightly. She wore a bright red form-fitting

In one of the few known photographs of Hoppity Hooper during his days at the Blue Moon Diner, he takes a much needed break from his demanding clientele. (Photograph courtesy of Halloway Bumpsteed, Jr.)

dress with a yellow zig-zag pattern. Her lipstick matched the red in her dress. Her brunette hair framed her slightly doughy features in an obvious attempt to emulate Betty Grable's current hair-do. "She says that drinking in the morning is the first sign of alcoholism."

"Hilda's an expert on alcoholism," said Hoppity, talking over his shoulder as he cracked eggs with one hand and poured pancake batter with the other. Tiny beads of sweat stood out in the hairs on the back of his crew-cut head.

"She should be," said Lansberg, braying like a donkey at his own joke. His Adam's apple bobbed up in down in time with his laughter. "She's been practicing for twenty years!"

This statement provoked more juvenile laughter from the crowd at the counter.

"Hmmmm!" murmured Blakeley. "That's very interesting. But you see, I've got a problem. I came in late last night, and-"

"Somebody stole your tires!" said all the kids at the counter in unison. They erupted in another chorus of braying laughter.

"That's right!" exclaimed Blakeley in surprise. "But how did you-"

"Happens around here all the time," said Hoppity. He hacked loudly and spit on the griddle, sending up a tendril of white steam which was imme-diately pulled up by the exhaust fan.

"Did you check your gasoline?" asked Lansberg. Everybody laughed again.

"Yep!" said the girl with the Betty Grable hair. "If they got your tires, you can bet they got your gas too!" She smiled brightly at Blakeley.

"Well, this is outrageous!" said Blakeley. "Why doesn't somebody report this to the proper authorities?"

"City cops don't come out here," said Hoppity. "They say it's out of their prediction."

"You mean, out of their jurisdiction?" asked Blakeley, a look of puzzlement coming over his face.

"Beats me, bub," said Hoppity, shrugging his shoulders indifferently. "They don't come out. Capiche?"

"Well, what about the county?" Blakeley asked indignantly. "Don't you have a county sheriff?"

Everyone laughed again.

"You mean Homer?" chortled one of the boys, a skinny buck-toothed kid with terminal freckles. "Homer never comes out here!"

"If he wasn't too lazy to come out here, he'd be too scared!" chimed in Lansberg.

The whole group had another good guffaw, this time at Homer's expense.

"Well, I've still got to find some wheels and tires," said Blakeley, now resigned to his predicament. "Is

In this painting by noted aviation artist H. Lowell Bumgartner, Zanesville School of Aviation Maintenance & Beauty Technology students are shown learning their trade on two of the Stinson 10-A aircraft utilized by the school for training. (Painting by H. Lowell Bumgartner, from the Whitley Speale Collection of the Bone Lake Research Museum)

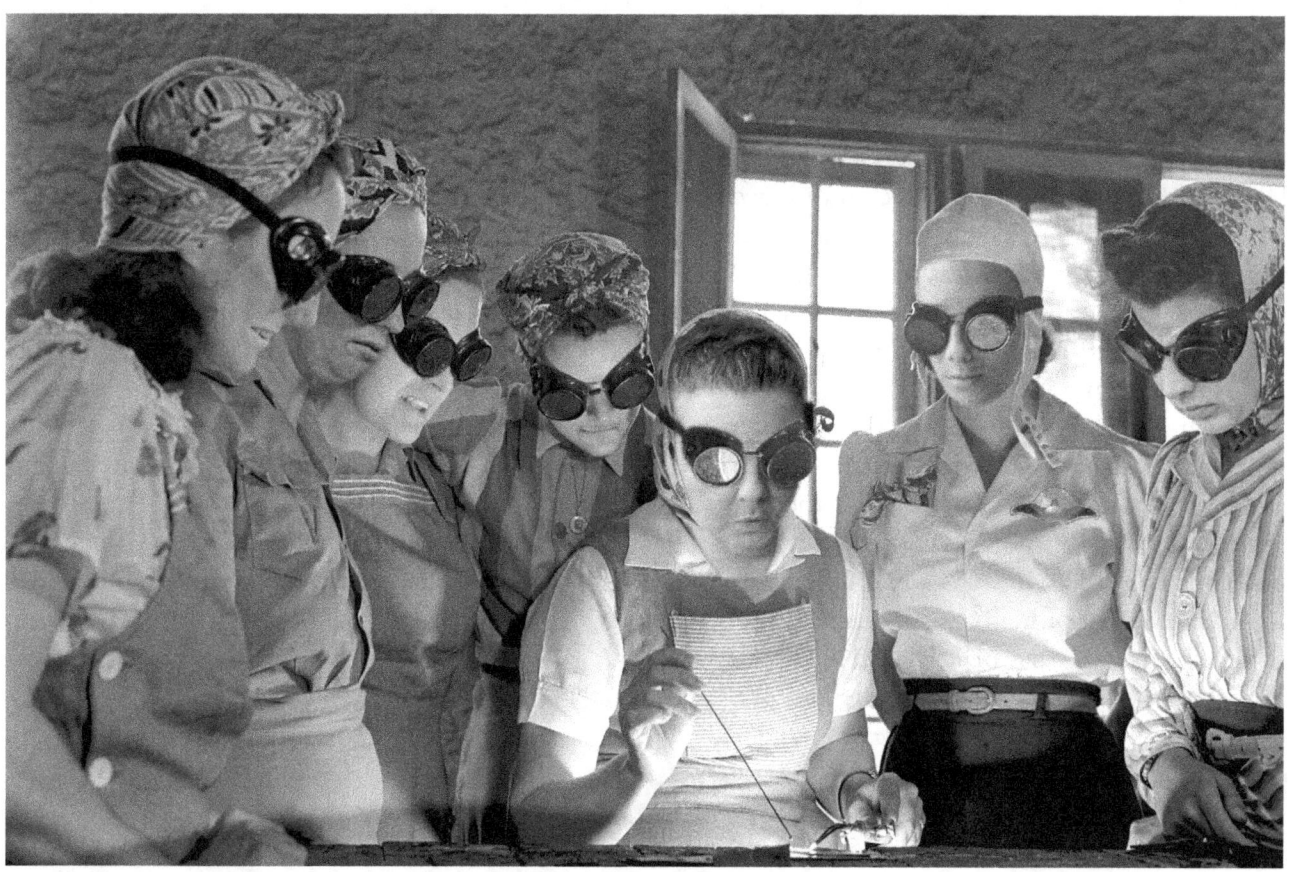

It wasn't all glitz and glamour for the beauty students at Z.S.ofA.M.&B.T. With the war raging, many of the beauty school students took aviation maintenance courses in addition to their beauty curriculum to better prepare them for service to their nation in time of need. Here Hilda Grackle, center, demonstrates the proper technique for gas welding. To her left, in the white flying cap, is Laurel Bergenstraum, receptionist. Third from the left is Amanda Smilch (neé Gunderson), cigarette girl for the KittyKat Top Hat Club, and certified Flight Beautician. Dr. Bothomfieder posits that in all likelihood, this photograph was staged for advertising and publicity purposes. Although Ms. Grackle was Chairman of the Board of Regents of Z.S.ofA.M.&B.T., it is highly doubtful that she ever taught a welding class there. (Photograph from the private collection of Dr. Eustis Bothomfieder)

there a service station close by?"

"Closest one's in town," said Hoppity. "That's about five miles as the crow flies."

"Does the crow come out here?" asked Blakeley. Everyone stared at him blankly.

"How do you get into town?" asked Blakeley.

"Well, there's a taxi in town," said the girl with Betty Grable hair. She smiled her ample lipped smile at Blakeley.

"But it doesn't come out here!" said everyone in unison. This cracked them up again, and they all cackled, slapping each other on their backs, as if in congratulations for their mutually keen wit.

"Why don'tcha try across the street to the School?" asked a lanky horse-faced brunette in a teal chinese print dress. On the fabric of the dress, coolies hoed in rice fields, carried water in buckets hanging from poles across their shoulders, and walked along beside two-wheeled carts pulled by water buffalo.

"Hey, yeah!" chortled the buck-toothed boy with all of the freckles. "They've always got tires!"

Everyone laughed except Blakeley, who has somehow missed the humor in the statement.

Blakeley looked across the highway. Standing in the middle of what looked like an abandoned corn field was a group of dilapidated buildings with a weather-beaten sign out front that read, "Zanesville School of Aviation Maintenance & Beauty Technology."

"Yes, well, thanks," said Blakeley. "I'll check over there. Over 'to the school.'"

He opened the creaking door and turned to leave. Behind him, the buzz of conversation immediately started up again as if it had never stopped.

"Sure thing, bub!" said Hoppity. "Come back when you got some money!"

LT. BLAKELEY'S TRADE SCHOOL ORIENTATION

Blakeley went back to his room and dressed. He walked back to the diner, which was now empty except for the chef. Evidently, recess was over. After a breakfast which included hash-browns upon which the pepper looked suspiciously like cigar ash, Blakeley got up from the counter. He closed the creaking door behind him, and crossed the dusty grey two lane highway. Blakeley walked up the crumbly sand-stone steps, and entered the front door of a clapboard house that hadn't been painted since the last war. Sitting sideways behind a reception desk, filing her inch long ruby red nails, was a girl who looked like she wasn't yet out of high school. She was not one of the girls that Blakeley had seen in the diner earlier. Evidently, she ate her breakfast at home. The girl was dressed in blue jeans rolled up at the bottom, and a sloppy Joe sweater, but incongruously, she sported the latest hair fashion. She blew a huge pink bubble with her gum and smacked it, a sound like a pistol going off.

"Ooooooh, a soldier boy!" she cooed. "What are you doing here, handsome?"

"Do you have a service station here?" he asked. "I'm staying at the motel across the street, and-"

"Somebody stole your tires?" she asked sweetly, batting her considerable eyelashes.

"Yes," said Blakeley. "As a matter of fact, someone stole my tires last night. How did you know?"

"You want to talk to Leo over in the Tech School," she said, and pointed to the hangar across the ramp. "It's in that building right over there. You can go into the door on that little side building."

"Thanks," said Blakeley, as he turned and walked to the door. He grabbed the handle, stopped for a moment, then turned back to the girl. How did you know-"

"About your tires? Oh, that happens to everybody who stays over at the Blue Moon!" The girl smiled brightly at Blakeley. "Come back and see me sometime!"

Blakeley walked toward a lean-to building attached to the side of the hangar. Even from across the ramp, he could hear a steady and raucous pounding emanating from the building. He entered a side-door, and the pounding doubled in intensity. He put his hands on his ears. Standing at a long table on the far wall, were about 20 teenage boys in sky blue coveralls. He recognized some of them as the same boys who were in the diner earlier. Each one was ferociously pounding on some unfortunate aircraft component with a huge ball peen hammer.

Blakeley looked around the room. Slouching against the wall opposite the boys was an individual

Hilda Grackle, or Miss Kitty Kat, as she was less formally known, in all her finery. This photograph presents Ms. Grackle as she was most often encountered at the entry way to the Kitty Kat Top Hat Club. (Photo courtesy of Dr. E.C. Whimpington)

that, in Blakeley's mind, could only be a juvenile delinquent, although he had never actually seen one before. The boy appeared to be only slightly older than the pounders, and was about Blakeley's size and build. He wore the same coveralls as the boys, only his were rolled down around his waist and the arms were tied in front. He wore a white T-shirt, with a pack of smokes rolled up in the sleeve. A cigarette dangled from his sullen lips. He was lean, with taut white skin stretched over the stringy muscles of his arms. A thick shock of dark greasy hair jutted forward above his head like the prow of a ship, hanging

precariously over his forehead before curving back over itself. He would have looked menacing, except for the black rimmed coke bottle glasses safety wired together at the nose bridge.

Blakeley walked over to the delinquent. "Excuse me," he yelled. "Who's in charge here?"

The delinquent looked up at him, and his cigarette bobbed up and down, indicating that his lips were moving. But Blakeley could hear nothing.

"I can't hear you!" Blakeley mouthed. The delinquent took Blakeley by the elbow and guided him out the door, shutting the door behind him.

Once outside, the din was only slightly reduced. Blakeley yelled again, "I said, who is in charge here?"

The delinquent frowned, held up one finger, and opened the door. Then he yelled inside, with a voice all out of proportion to the slight body that it emanated from. It was like a bull gorilla bellowing in the jungle.

"Heeyyy!! Knock it off, will yas? I'm tryin' to talk heah!"

Immediately the din stopped. The delinquent turned back to Blakeley and said, "Hammer practice. I hate it. Now, whaddya want, Mac?"

Blakeley didn't answer. The delinquent reached up and grabbed Blakeley's wrists, pulling his hands down from his ears. "I said, 'Whaddya ya want, Mac?'"

Blakeley shook his head. "What's going on in there?"

"Calibration class. You know, with hammers," The delinquent's lips moved upward in a reasonable semblance of a smile. "I'm Leo, the instructor. Say, you a sailor?"

"Do you have a service station here?" Blakeley asked, ignoring the question.

"I'd be in the Navy myself, if it wasn't for my record," said Leo. "Natch, I'm doin' my part here, ain't I. Teaching these Bozos all about calibration."

His lips moved upward at the corners again, twice in rapid succession. Blakeley got the impression that he was pushing a button to move them.

"Say, do you think you could pull some

At the height of the depression, Z.S.ofA.M.&B.T. utilized federal funding to create publicity shots of its students at various work activities in order to enhance enrollment. The program's effectiveness was nil, and enrollment continued to decline, but the project provided a unique record of the students. Here, Carolina Gundersen, older sister of Amanda Smilch (neé Gundersen) is shown operating a drill press at Prestley Industries in Columbus, Ohio. Carolina, like so many other Z.S.ofA.M.&B.T. students, interned at Prestley Industries doing subcontract work for Brewster Aeronautical Corporation. Professor Flungk collected a hefty fee for providing the interns, who, although unpaid, were only too happy to pay tuition for the privilege of working for free. (Photo courtesy of Dr. E.C. Whimpington)

strings and get me into the Navy? I wouldn't have to stay or nothin', maybe just long enough to at least get me a uniform? The dames love those uniforms." He shook his head sadly. "I just don't get it."

Although unable to talk Blakeley out of a uniform, Leo was eventually able to piece together enough components to pass himself off as "Raoul" at Miss Kitty Kat's Top Hat Club. He never wore his "uniform" to work, preferring his trademark t-shirt and blue jeans. Likewise, "Raoul" never wore his coke bottle glasses to the Club, as they would have raised questions about his fitness to be a member of the Free French Air Force. (Photograph courtesy of the Whitley Speale Collection of the Bone Lake Research Museum)

"Somebody stole the wheels and tires off my car last night." Blakeley said, patiently. "I need to get them replaced."

Leo frowned. "Ah, criminy, fella, that's too bad. The nerve of some guys! What kind of car you got?"

"It's a '42 Ford sedan." Blakeley answered.

Leo's eyes lit up, genuinely this time. "Heyyyy!! A '42 Ford? Only you Navy guys could get a brand new car! You got a flathead V-8 in there? If a guy had a flathead V-8, he could outrun any cop in the county!"

"I don't know, I guess it might be," Blakeley shrugged his shoulders. "It's a motor pool car."

"Tell you what," Leo said. "You gotta be the luckiest guy in the world. Thing is, we got this order, for a special customer. But you bein' a sailor and all, I tell you what I'm gonna do. I'm gonna make you a deal. And guess what this special order is?" Leo punched Blakeley playfully in the arm. "Four tires mounted on Ford rims! Don't that knock your socks off? Who'd of thunk it? Not in a million years, huh!"

"Hmm," murmured Blakeley. "That is a coincidence."

"Come right in here!" Leo grabbed Blakeley by the elbow again and guided him around the side building and into the front of the hangar. "Be careful. They're still wet. They just came out of the paint shop this morning."

Leaning against the hangar wall were four shiny black tires on fire engine red rims. As Blakeley gaped at them, a fly orbited around one of the tires. It landed on the rim, and was stuck fast in the wet paint, its wings buzzing frantically. Blakeley stood one of the tires up. He twirled it around. The unpainted back side of the rim was olive drab, the same color as his service Ford.

"Like I said, they was for a special customer, but hey, I'll let you have 'em all four for a C-note," Leo said. "I'm losin' money, what with the new paint job and all that. What the heck, it's for the war effort, right?"

Leo punched Blakeley in the arm once again, but Blakeley didn't respond.

"Say, you got a spare tire? We can paint it to match," asked Leo. "Hey Mac! You all right, Mac?"

The "Boomerang" concept, as envisioned by Professor Hermann Flungk on that fateful New Year's Eve of 1934, at Miss Kitty Kat's Top Hat Club, bore little resemblance to the aircraft that he eventually would produce. (See page opposite. Artifact courtesy of the Whitley Speale Collection of the Bone Lake Research Museum)

THE PROFESSOR'S LEGACY

Blakeley didn't answer. He was transfixed, as if in a daze, his eyes on a fire-engine red and silver contraption that took up the right rear quarter of the hangar. Finally, he regained his voice. Pointing to the monstrosity, he asked, "What is. . . . that?"

"I'll make 'em to you for eighty bucks." Leo said, and added quickly, "And a sailor suit. Like yours."

"No, no, what is that. . . thing? That red and silver thing over there in the corner," Blakeley asks once again.

Leo held his arm up against Blakeley's arm. "We're about the same size, wouldn't you say?"

Leo pressed his smile button again, and his lips moved up and down at the corners. "We could just trade clothes!"

Blakeley grabbed Leo by both arms, and shoved him against the hangar wall. "Is that yours? Did you build it?"

"Criminy, Commodore, you'll crush my smokes!" Leo whined. "Take your mitts off me, I ain't done nothin' to you!"

Blakeley let go, and Leo rubbed his arms

"That's the Professor's big project," he said. "His 'Flying Boomerang.' Yeah, he's the one thought it up alright, but I'm the one got it built. You try gettin' any real work out of these high school dropouts. Ain't easy, I'll tell you."

Like a sleepwalking dreamer, Blakeley crossed the hangar to the ungainly aircraft, if indeed that was what it was. Coming closer, he saw that the aircraft, or what was left of it, had clearly been wrecked. The left engine was canted down at an unnatural angle and the wooden propeller had been splintered. The right engine appeared to be missing. The tail section had been hacked off and was standing in the corner behind the airplane.

Blakeley walked up to the front of the strange craft. It was covered with a thick layer of dust and bird droppings. He wiped his hand across the dusty fuselage of the strange contraption, sending a flurry of dust speckles into the still air of the hangar. Where the dust had been wiped away, the mystery aircraft was painted a bright fire engine red. Blakeley wryly noted that the red paint on this apparition was the same paint as the fresh paint on the four automobile rims.

Blakeley ran his fingers slowly over the hand

painted fuselage, then out across the rusty corrugation of the starboard wing. He felt like an Egyptologist entering a wealthy Pharaoh's tomb for the first time. He gazed wonderingly at this forlorn and forgotten artifact, trying to grasp it in its entirety.

The craft certainly appeared to have a fuselage. And wings. The left engine was a type that Blakeley, a seasoned mechanic, had never seen before. It was an inline design, of at least twelve cylinders. He counted the exhaust ports. Fourteen! Such a huge engine, Blakeley thought, must have been very powerful.

He examined the engine further. Sitting on top of the boxy cowling is what appeared to be a shiny brass spittoon. Wonderstruck, Blakeley slowly walked around to the ungainly nose of the aircraft. He stared in amazement at the right wing. The right engine was not missing; it was mounted backwards!

Blakeley was at a loss as to the purpose of this strange machine. Perhaps it was a test vehicle, designed to explore the flight characteristics of a pusher engine configuration. But if that were the case, then why wouldn't both engines be mounted as pushers?

Blakeley stepped back from the craft to get a wider view. Mounted on top of the fuselage was what appeared to be a helmet mounted on a pole directly behind the cockpit. It was obviously designed to fit over the pilot's head, but to what purpose? Perhaps it was a radio receiver, or some sort of protective headgear. Blakeley could only stare in wonder at the aerial perplexity. Stunned, he turned, striding purposefully back across the hangar floor to where Leo stood, sulking.

"Who is this professor, the man who designed this machine?" Blakeley demanded. "Where do I find him?"

Leo pouted, rubbing his arms where Blakeley had grabbed him earlier.

"Ya din't have to grab my arms like that!" Leo sulked.

"Damn it, man, this is of the utmost importance! The outcome of this war may well depend on it!" Blakeley pointed at the pathetic contraption in the corner. "Who designed that aircraft?"

"How would you feel if someone grabbed your arms like that?" pouted Leo.

Frustrated, Blakeley strode out the side door of the hangar.

One of the few surviving photographs of the Flungk Z-44 Mark I. An unknown photographer had taken this shot of itinerant agricultural Stearman biplanes which happened to be using the airfield for local dusting and spraying operations. The Flungk "Boomerang" can be seen sitting forlornly in the background, upper right. (Photograph courtesy of Halloway Bumpsteed, Jr.)

"Hey!" Leo yelled at the retreating figure of Lieutenant Blakeley. "Come back here! You want those wheels and tires, or not?"

The receptionist who confronted a frustrated Lieutenant Blakeley in the offices of Z.S. of A.M.&B.T. In actuality, her name was Laurel Lautermilk, and she would figure prominently in the saga of the Flungk "Boomerang." (Photograph courtesy of Halloway Bumpsteed, Jr.)

BLAKELEY PENETRATES THE VEIL

Blakeley strode purposefully across the broken tarmac, back to the main office.

Behind Blakeley, in the hangar, Leo yelled, "What are you numbskulls standin' around for, anyhow? Get back to work!"

Immediately the hammering resumed, and Blakeley put his hands over his ears once again. Walking up to the dilapidated office, he tried to open the door with his elbows. This was unsuccessful. Wincing, he reluctantly uncovered his ears, opened the door and rushed into the office.

The receptionist still sat behind the desk, only now she was busily applying mascara to her eyelashes, gazing into a pink rhinestone studded compact that was shaped like a clamshell.

Blakeley walked up to the desk and waited, obviously fidgeting. The receptionist, engrossed in her make-up mirror, ignored him.

"Ahem! Excuse me, Miss!" Blakeley yelled, straining to be heard above the distant hammering.

The receptionist, startled, peaked around the compact mirror. She saw Blakeley and smiled. She sat down the compact mirror. Making a closing motion with her hands, she pointed to the open door behind Blakeley. He closed the door. The din from the hammering was noticeably subdued.

"It's the soldier boy!" the receptionist smiled brightly. "I told you to come back and see me, but I didn't know it would be so soon!"

"Who's in charge here?" Blakeley asked gruffly.

The girl looked puzzled. "Gee, fella," she said, "Like I already told ya'! Go out to the hangar!"

"No!" shouted Blakeley, growing more frustrated. "I mean, who's really in charge? Of the entire school?"

The receptionist smiled brightly again "Ooooh!" she said. "You mean, like who's in charge of everything? Like the Professor?"

"Yes, yes!" said Blakeley. "Like the Professor! Where can I find him? Where is he?"

The girl looked puzzled again.

"Well, sheesh, Mister," said the girl. "Ain't he in his office, like he always is?"

"How would I know if he is in his office or not?" Blakeley asked, exasperated. "It is vital that I speak with him! Now!"

"Oh, you mean, you want to see him?" asked the girl.

"Yes, immediately!" said Blakeley. "Now! This instant!"

"Well, gee, that's too bad, mister," said the receptionist, returning to her make-up. "Because nobody gets to see the Professor without an appointment." She deftly applied more mascara to her already huge eyelashes.

"Well, who do I see about an appointment?" Blakeley asked.

"That would be me," said the receptionist, putting away her mascara. She looked at herself in the mirror, raising one eyebrow appraisingly.

Blakeley waited. The girl smiled at herself approvingly.

"Well?" Blakeley asked finally.

The receptionist looked up from her mirror again, smiling sweetly.

"An appointment!" yelled Blakeley. "Would you please make me an appointment! Now!"

"Well, my stars!" exclaimed the receptionist. "You can't just walk in here and make an appointment just like that!" She snapped her fingers in the air. "First, I'd have to check the Professor's calendar and see if he was available."

She reached into her purse, and pulled out a brass lipstick tube. Making an O with her lips, she drew a greasy red line around her puckered mouth.

"So?" screamed Blakeley. "Check his calendar!"

The receptionist smacked her lips together three times, making a little popping sound each time.

"I'm sorry, mister," she said sweetly. "I would certainly like to help you, but I can't check his calendar right now."

Blakeley stood rigid, his arms straight at his sides, fists clenched.

"Okay," he fumed through gritted teeth, "Why can't you check his calendar?"

"Because I'm on my break now," she smiled, shrugging her shoulders.

"Young lady, I'll have you know that my inquiry concerns a matter of the utmost importance to the defense of this nation! Need I remind you that we are at this very moment in a state of war-"

"Oooh!" squealed the receptionist. "Did you know that when you get worked up like that, your left eyebrow arches up? It makes you look just like Clark Gable!"

Blakeley, halting in mid-sentence, smiled sheepishly.

"Really?" he asked.

She looked at him for a minute, frowning.

"Naah!" she said dismissively, as she began filing her nails. "It looked like Clark Gable's eyebrow for just a second, but it's gone now."

"Yeah, but for just a second, like you said . . . " said Blakeley.

The receptionist gazed at his face for a long moment, scrutinizing his features.

"Naah!" she said, returning to her nails.

BLAKELEY TRIES TO PLAY THE SECURITY CARD

"Hmph!" Blakeley frowned. "Young lady, whether you are aware of it or not, this is serious business. National security is at stake! If I have to contact the local authorities, let me assure you that I will not hesitate-"

"Hah!" She looked over her left shoulder. "Hey, Della!" she yelled.

A girl stuck her head out from the door behind her. It was the girl from the diner with Betty Grable hair. She blew a pink bubble that exploded with a loud smack. She flicked the exploded gum off her lips with her tongue.

"Yeah?" Della asked.

The receptionist pointed at Blakeley with her nail file.

"The soldier boy's gonna call the cops," said the receptionist brightly. "He thinks they'll come out here!"

Both girls laughed uproariously.

"I don't have to fool around with your local yokel cops!" ranted Blakeley. "I've got contacts in Washington. Why, I can have the FBI down here in-"

"How're you gonna do that? Mister 'I don't have any tires on my car'!" The receptionist expertly mimicked Blakeley's slightly nasal tone, and both girls broke out in laughter again.

"I don't need a car!" blustered Blakeley. "Why, with just one phone call, I-"

The receptionist tapped the big black telephone on her desk with her nail file. "This school has the only telephone line this side of the city."

"So what if it is?" stormed Blakeley.

The receptionist tapped her nail file on the metal desk top.

"Now, let me get this straight," she said."You want to use my telephone to report me to the FBI?"

She looked over her shoulder once more at the girl with Betty Grable hair.

"Ain't he a scream?" she asked.

Both girls howled with laughter.

Then the girl with Betty Grable hair slapped her hand through the air.

"You two crack me up!" she said, between fits of laughter. Then she turned and walked back into her office.

Della Bergenstraum, the "girl with the Betty Grable hair," wasn't above helping out in the aviation maintenance shop as the need arose. (Photo courtesy of Dr. E.C. Whimpington)

Blakeley boiled silently for a moment. Then, he tried another tack.

"Look," he said earnestly. "Can't you help a guy out? This is really important. I'm sorry if I got a little . . . upset. It's just that I really need to see Professor . . . What's his name?"

"Flungk."

"I beg your pardon?"

"Flungk. That's his name. Professor Flungk."

"Okay," said Blakeley. "Professor Flungk. Yes."

"Professor Hermann Flungk."

"Professor Hermann Flungk. Yes. I see. Well, it is vitally important that I see Professor Hermann Flungk. The fate of the United States may very well depend on it! You have no idea—"

"Whoops!" The receptionist held up both hands

in front of her, palms outward. "There you go again!"

"I'm terribly sorry!" said Blakeley, wringing his hands. "What I mean is, is there any possible way that you could possibly check the Professor's schedule, and see where you might possibly fit me in? If that's possible, that is."

The receptionist closed her compact mirror, put the cap on her lipstick, and closed her nail file. She deposited them all in a tan purse that she closed with a loud snap.

"Well, now that my break is over, I suppose I can see what I can do," she said. She folded her hands in front of her and smiled brightly. "Now, how may I help you?"

"I just told you!" sputtered Blakeley.

"Just kidding!" said the receptionist.

"No, perhaps you are right," said Blakeley solemnly. "Maybe it is best if we start over. My name is Lieutenant Julian Blakeley and I—"

"How do you spell that?" asked the receptionist, pulling a form out from a side drawer and inserting it into her typewriter.

"Capital B, l, a, k—"

"No, Julian," said the girl. "How do you spell 'Julian'?"

"Capital J, u, l, i—"

"Isn't that kind of a sissy name for a soldier?" she asked. "What do your friends call you? Hank? Butch?"

"They call me Julian, and really, what does this have to do with—"

"I just thought a soldier would have a tougher name, that's all. I went steady with a guy named Maurice in high school, but he went in the Navy, and now he's Dirk."

"If you could just check the schedule—"

85

An extremely rare photograph of Professor Hermann Flungk. Virtually a recluse, with the exception of his nightly forays to Miss Kitty Kat's Top Hat Club, the Professor rarely left his office, or allowed any visitors. This photograph was taken on the occasion of the demonstration of a remote flying control for the Z-44 "Boomerang," which would eliminate the need for a pilot. However, as it was subsequently demonstrated that the Z-44 was uncontrollable, there was little demand for the device, and production languished. (Photograph courtesy of the Whitley Speale Collection of the Bone Lake Research Museum)

She sighed dramatically as she opened the second drawer and pulled out an ancient leather backed ledger.

"Just a moment, while I check," she said officiously.

Blakeley fidgeted as she leafed through the pages.

"Hmmmmm!" she murmured. Abruptly, she slammed the ledger shut. "I'm sorry, there doesn't seem to be an opening."

"Do you honestly expect me to believe that this man, this Professor, is so busy that he doesn't have any time for appointments? What does he do all day?" demanded Blakeley.

"Nobody knows," said the receptionist. "He's very secretive. He never opens his office door."

"How do you know he's really in there?" asked Blakeley.

"Every so often, you can hear a flush," said the receptionist. "He has his own toilet."

"If he's not doing anything, why can't he see me?" screamed Blakeley.

"I suppose he could, if he wanted to," said the receptionist. "He just doesn't want to, is all."

"How do you know," yelled Blakeley. "if you don't ask him?"

"Oh, I know," said the receptionist, as she put the big ledger back in the drawer. "You see, I have standing orders never to make an appointment for anyone with Professor Flungk."

"Then why did you tell me you could make an

Amanda Gundersen, left, and Della Bergenstraum, right, leave the Muskingum County Courthouse, probably in 1939. They were likely in attendance for one of the many competency trials of Professor Flungk, although it is unknown whether they were testifying for or against the Professor's sanity. (Photograph courtesy of the Whitley Speale Collection of the Bone Lake Research Museum)

talk to him. Spreichen ze Deutche?"

"I beg your pardon?" said Blakeley.

"Deutche! German! Kraut talk! Can you speak it?" asked the receptionist.

"I had two semesters of French in high school," said Blakeley.

"The Professor only speaks German," said the receptionist calmly. "Even if you did get to see him, you couldn't understand him. And he sure couldn't understand you."

"Well, what if I wanted to enroll in Beauty College?" asked Blakeley.

"You could talk to Della, here," said the receptionist.

Della flipped her Betty Grable hairdo and winked at Blakeley, holding her chubby hand up by her cheek and waving to him.

"Are you considering a career change?" Della asked.

Blakeley ignored her. "What about the Aviation Maintenance courses? Who would I see about those?"

"Let me see," said the receptionist, as she opened the drawer and pulled out the big leather backed ledger again.

She leafed through the pages, running her fingers down the columns until Blakeley was about to explode.

"I think I could arrange for you to meet with Mr. Hooper in late February. How about the twenty sixth? Would you prefer morning or afternoon?"

"Who is Mr. Hooper? Would he know anything about that machine out in the hangar?" asked Blakeley.

"Mister Leo Hooper is Professor Flungk's teaching assistant for Aviation Maintenance Technology. Perhaps he could help you-"

"Leo?" screamed Blakeley. "Leo? That moron in the hangar? I just got away from Leo! He's a nincompoop! A miscreant! A malingerer! He's incompetent!"

"Well, I'm sorry I couldn't help you," said the receptionist, frostily. "Obviously, we don't measure up to your high standards. Perhaps if you inquired at another institute of higher education-"

appointment for me with him?" screamed Blakeley.

"Oh, no, I just told you that I'm the one to see about getting an appointment. I didn't tell you I could make an appointment with the Professor for you. You made that up all by your little lonesome!"

Della poked her head out of the door and pointed her pencil at Blakeley.

"That's right!" Della said triumphantly. "She never told you she could do that!"

"Ummmppphhhh!" screamed Blakeley.

"Do you see how you're just getting all worked up over nothing?" the receptionist asked. "Besides, even if you did get to see the Professor, you couldn't

But Blakeley didn't hear her. He was already out the office door, running back across the broken tarmac to the hangar.

It was much quieter now. Hammer and calibration class was evidently over.

BLAKELEY'S LAME ATTEMPT TO PUT IN THE FIX

Blakeley looked frantically around for any sign of Leo, but he was nowhere to be seen. He opened the side door to the lean-to building, and stepped inside. From the other end of the hall he could hear voices, punctuated by loud guffaws.

Blakeley ran down the hallway. White smoke was rolling out of the doorway of the room at the end. He turned the corner and looked inside. There was a bare wooden table and some folding chairs. Leo was leaning against a battered red Coke machine, pontificating to the students of the calibration class, who were obviously on a smoke break.

"So I says to him, heh, heh, I says to him, 'Hey, you know who you're talkin' to?" Leo unrolled a pack of Chesterfields from the arm of his T-shirt and smacked them on the wooden table. He patted one out of the pack and lighted up, inhaling deeply.

Blakeley entered the room tentatively, advancing toward Leo. "Excuse me," he said. "Leo? You are Leo, aren't you?"

Leo brightened. "Hey, fellas, this is the guy I was just tellin' ya about! Where'd you go, Commodore? You kinda disappeared on me back there!" He frowned. "Say, if you're lookin' for those tires and wheels, I got some other guys lookin' at them now, so the price has gone up!"

"Are you Leo Hooper, the Professor's teaching assistant for Aviation Maintenance?"

"Am I Leo Hooper? Am I Leo Hooper? Does Lucky Strike make fine tobacco?" Leo mugged for the boys, who howled with laughter.

"Here's twenty dollars," said Blakeley, walking up to Leo. "It's yours if you can get me in to see Professor Flungk-"

"Whoa, whoa, whoa!" said Leo, grabbing the twenty ones and stuffing them into Blakeley's shirt pocket. "Let's go down to my office where we can talk in private, Commodore. Excuse us for a few minutes, won't you, boys?"

Leo put his arm around Blakeley's shoulder and guided him hurriedly out of the room. He took Blakeley down to the other end of the hall.

"Commodore, don't you ever wave that kind of lettuce in front of those mokes back there! Why, there isn't one of them who wouldn't carve your guts out and leave them layin' on the sidewalk for just one of those greenbacks there!" He patted Blakeley's pocket fondly.

Blakeley looked back down the hall. Three heads were popping around the door frame, watching them.

"Yes," Blakeley said. "Thanks for the tip. Now, can you help me? I really must speak with Professor Flungk."

"Well, Commodore, it ain't as easy as it sounds. The Professor, you know, he's kind of . . . funny, you know? I mean, he doesn't just talk to anybody."

"Does he talk to you?"

"Does he talk to me? You're kiddin' me, right? Does the Pope talk to the big man upstairs?" Leo slapped Blakeley on the back, grabbed the twenty bucks out of Blakeley's shirt pocket and stuffed it in his own pants pocket.

"The Professor's a busy man, what with all the thinkin' he's got to do, and in a foreign language to boot!" Leo said. "But I'll see if I can't put in a good word for ya! Now, about them tires . . . . ."

10. BLAKELEY MEETS MISS KITTY KAT

A German Heinkel HE-111 over England during the Battle of Britain, with open bomb bay doors. (Royal Air Force Battle of Britain campaign diaries)

HILDA DOES HER BIT FOR THE WAR EFFORT

(Excerpt from *The Flight of the Boomerang*, by Elmer C. Wackmallit, continued.)

After Blakeley had purchased the tires from Leo and arranged for their installation, he went back to the Blue Moon Motor Inn. Hilda sat on a stool behind the counter, idly tapping a cigarette in a brass ashtray.

She was reading a "Secret Lives of the Stars" magazine. She wore a faux ermine housecoat, blue fluffy slippers, and a leopard print shower cap on her head. She looked up as Blakeley entered.

ADDENDUM: THE CONCEPT OF STRATEGIC BOMBING AND ITS RELATION TO THE DEVELOPMENT OF THE "FLYING BOOMERANG"

The concept of bombing is almost as old as the airplane itself. During World War I, airplanes were first used for aerial surveillance. They were considered a longer range version of the stationary observation balloons which had been in service since the U.S. Civil War. However, the temptation to drop something from an airplane proved too

89

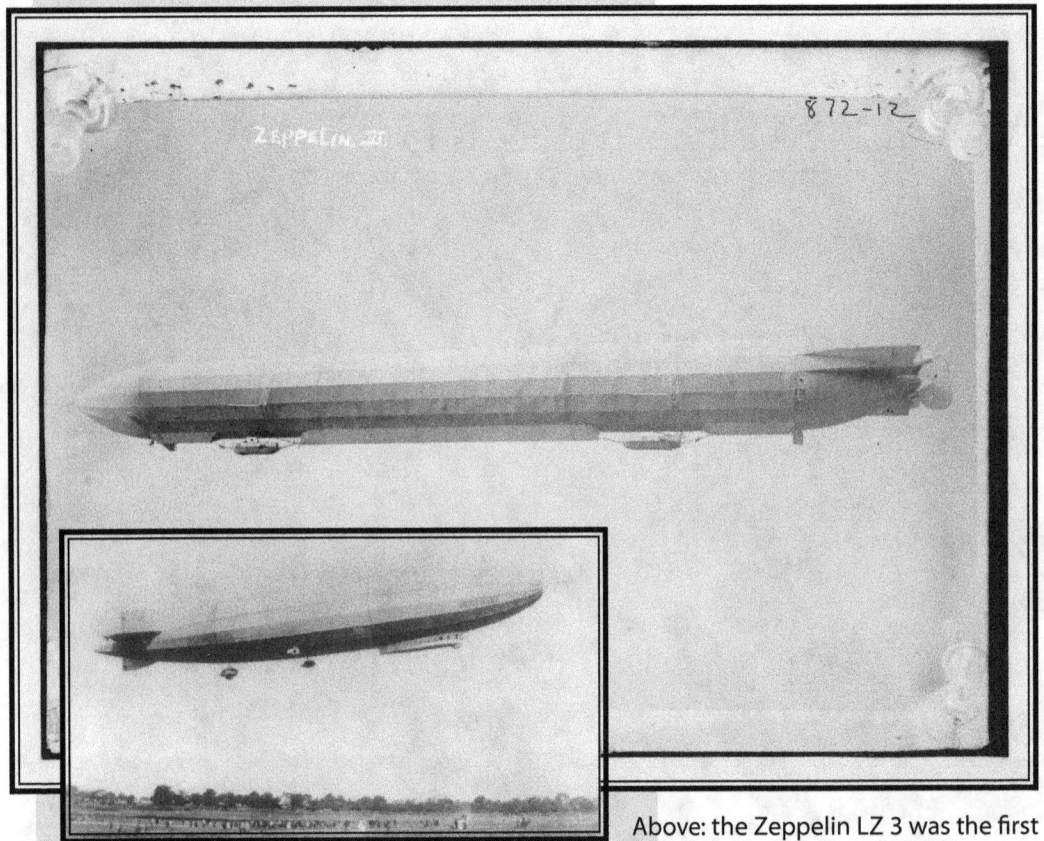

Above: the Zeppelin LZ 3 was the first truly successful Zeppelin design. It was first flown on October 9, 1909. After its military acceptance trials, it was renamed the Zeppelin Z1, and remained in service until March 1913, when it was retired as obsolete. Note the straight line, unstreamlined fuselage, the open gondolas, and the arcane and overly intricate tail empennage, as compared to the later Zeppelin "Bodensee," (inset).

"Hello, sugar," said Hilda. "You're that fellow looking for the Professor, right?"

Blakeley, who was nodding politely as he walked past Hilda to his room, stopped in his tracks. He shook his head in bewilderment.

"What is it about this place?" Blakeley looked up at the ceiling, as if beseeching an answer from the tawdry lobby chandelier. "Why does everyone here seem to know my business even before I do?"

"No mystery to that, sugar," said Hilda. "We don't get that many visitors, especially from the armed services. Makes you kind of . . . interesting, that's all."

Blakeley sighed and walked up to the counter. "I might as well settle up with you for another night. Looks like I'll be sticking around for at least that long."

"You're darned straight about that," said Hilda. "Excuse my French. But it's already three o'clock, you're still here, and check-out is at noon."

"That's all very well," blustered Blakeley. "But how am I supposed to check out when you don't get out of bed till two o'clock?"

Hilda smiled tiredly, and took a long drag on her Pall Mall. She exhaled through her mouth and simultaneously inhaled the smoke through her nose. "Seems like you're not the only one around here whose business everybody knows."

She tapped her unfiltered cigarette in the ashtray, and set it down. Sticking out her tongue, she plucked a minute piece of tobacco from it, looked at it critically for a moment, then flicked it casually against

great. Soon, aircrews began carrying bombs, often constructed from artillery shells with fins added to drop on enemy troops and positions. Thus the concept of tactical bombing was born.

Tactical bombing is an aerial attack on an enemy position or personnel in order to provide an immediate benefit to friendly forces engaged in the conflict.

Strategic bombing, by contrast, is bombing for the purpose of destroying a nation state's ability to wage war. These bombings would characteristically occur over a longer period of time, perhaps weeks or months. The goal is to accomplish the attacker's objective through the destruction of wartime production and transportation capability (armaments and munitions factories, railroad yards, oil fields, etc.), or through the destruction of the enemy's will to resist by attacking the civilian populace

The Martin MB-2, above, was the state-of-the-art bomber when Billy Mitchell demonstrated the efficacy of airpower against the battleship in July, 1921. In the inset, upper right, an MB-2 drops a 2000-pound bomb in an intentional "near miss" in order to damage the stern hull plates of the German battleship "Ostfriesland," causing her to sink. In the inset lower left, a Martin MB-2 scores a direct hit on the USS Alabama with a phosphorous bomb. In the lower right, a Chicago Tribune cartoon comments on the trials.

the back wall.

"Never mind," Hilda said. "I usually cut some slack on the last day. Just lock up your room when you leave and drop the key in that slot over there."

She pointed with her head at a lock box on the end of the counter.

"I mean, it's not like anybody would call the law on you or anything," said Hilda. "What's your name, sugar?"

"I'm Lieutenant Julian S. Blakeley, Air Corps," said Blakeley.

"I know that, sugar," said Hilda. "Rhonda, the night girl, told me after you checked in. Besides, that's all in the register. I just wanted to know what they call you, that's all."

Blakeley looks perplexed. "Well," he stammered. "Julian, I guess."

"Well, Julian, sugar, that will be six dollars,

in urban centers. In theory, these attacks would make the prospect of a negotiated peace or surrender more desirable than continued bombing.

It is outside the scope of this addendum to address the moral aspects of strategic bombing, a debate which has existed for as long as the strategic bombing concept, and remains contentious to this day.

Rather, this addendum is concerned solely with the development of strategic bombing and the effect, real or perceived, that these strategies exerted upon the conception and development of the Flungk Z-44 Mark I "Boomerang."

The first instances of strategic bombing, (although they were not called that at the time), occurred in World War I. Most notable were the German Zeppelin attacks on England from February 1915 to August 1918. At times, these raids caused considerable damage and terrorized the population.

After the First World War, General Billy Mitchell, a staunch proponent of airpower, advocated strongly for a strong and separate U.S. Air Force, and the development of Naval aircraft carriers. His spectacular sinking of the German battleship "Ostfriesland" by Martin NBS-1 bombers in static trials conducted by the Navy in July, 1921 presaged the end of the battleship era, and in many ways, marked the beginnings of modern air power. He was later brought up on court-martial charges for confronting his superiors on their refusal to pursue airpower in defense of the nation.

Later in the interwar period, the Spanish Civil War of 1937 to 1939 provided an opportunity for the fledgling Luftwaffe to test its air tactics and aircraft in actual battle conditions. The Condor Legion was dispatched by the Nazi regime to support the Fascist rebels of General Franco. Composed of German Junker JU-52 transports, Dornier DO-17 and Heinkel HE-111E bombers, Junkers JU-87 Stuka dive bombers, already obsolete Heinkel H-51 biplane fighters, and eventually BF-109 fighters, the Condor Legion foreshadowed the coming action of World War II, most notably in their bombing of the Basque Republican enclave of Guernica. The bombing of Guernica was ostensibly to destroy a railroad station and bridge used by the Republican forces, but the bombing activity, which continued for

plus tax, for one more night," said Hilda, blowing a perfect smoke ring. "My name's Hilda, by the way, but everybody calls me Miss Kitty Kat." She pointed with her head at the curtained entryway to her left. "Because of the Club, of course."

Blakeley pulled a money clip from his front pocket, and peeled off a five and two ones, laying them on the counter. "Keep the change," he said.

Hilda folded the bills and stuffed them in her bra. "Thanks," she said. "The folks that come through here usually aren't that generous."

"Have a nice day," Blakeley cut Hilda a crisp smile, and turned to go to his room.

"I was just wondering," said Hilda. Blakeley stopped and turned. "You didn't give any money to Leo, by any chance, did you?"

"Well, it certainly is a small world, isn't it?" said Blakeley.

"I don't mean to pry, sugar," said Hilda. "It's just that–"

Above: Condor Legion volunteers load the bomb bay of a HE-111-E during the Spanish Civil War. Note the stepped canopy of the E model, and the lack of a defensive machine gun in the nose bubble. Inset, left: HE-111-H models in flight in a 1940 German propaganda photograph. (Photographs courtesy of Deutsches Bundesarchiv)

Hilda blew out a long puff of smoke with a tired sigh. "Well, he's my sister Zelda's boy, Lord rest her poor soul," she said. "I don't want to talk bad about my kin, but that boy . . ."

Above: The Junkers JU-52 was originally built as a civilian transport, but the Condor Legion used 18 JU-52s equipped with bomb bay doors to bomb Guernica on April 26, 1937. Inset: The bombing resulted in the almost total devastation of the town. (Photographs courtesy of Deutsches Bundesarchiv)

"Well, I wouldn't worry about it," said Blakeley. "You see, someone stole my tires last night. Leo happened to have a set, so I bought those."

"I see," said Hilda, her eyes rolling to the ceiling. "Did he charge you extra for putting them on your car?"

Blakeley smiled. "You really needn't worry. I'm on official business. I'll get reimbursed for my expenses."

"As long as you're satisfied," said Hilda. "Did he sell you anything else?"

"Well, he-"

"He told you he's got an inside track with the Professor, right? Get you right in to see him, right? How much did that cost you?"

"I just assumed, since he's the Professor's teaching assistant, he-"

Hilda exhaled a huge puff of greyish Pall Mall

three hours, devastated the town. Thus was the pattern set that would be repeated throughout the coming war. Bombing missions would be initiated against targets of strategic military significance. Then, owing to faulty navigation, or adverse weather conditions, or intentionally, bombs would also fall on civilian targets. (In the attack on Guernica, the targets of the attacking Luftwaffe pilots were almost immediately obscured by smoke and dust. Nonetheless, the attack was pressed, resulting in the virtual destruction of the town, and the subsequent rage and controversy that enveloped the incident.)

On September 1, 1939, Nazi Germany invaded Poland, marking the beginning of the Second World War. The invasion also popularized the concept of Blitzkreig, or Lightning War, as a coordinated attack of armored and mechanized ground forces utilized in conjunction with close tactical air support, primarily in the form of the Junkers JU-87 "Stuka" dive bomber. While the Blitzkreig strategy may not have been an actual doctrine of the Nazi Wehrmacht, it defined the function of the Luftwaffe, as envisioned by the Nazi leadership, as one of close air support in cooperation with ground units, with a pronounced lack of emphasis on strategic bombing operations.

After the fall of France to German troops, Hitler turned his attention to Great Britain. The German Kriegsmarine had been decimated in the Norwegian campaign. Thus, the success of "Operation Sea Lion," the German plan for invasion of the British Isles, depended on the strength of the Luftwaffe to counter the British Royal Navy's mastery of the sea. But the Luftwaffe's relatively slow and clumsy JU-87 Stuka dive bombers could not operate effectively against the Royal Navy until air superiority over the Royal Air Force was achieved. The RAF, then in a seriously weakened condition due to the attrition of its pilots in the continental campaigns was all that stood in the way of Hitler's plan of conquest. At this point, Hitler and Goering were now committed to an air war to eliminate the RAF fighter force.

Throughout the summer and autumn of 1940, the Luftwaffe relentlessly attacked the British forces, first utilizing the JU-87 Stukas that had served them so well on the continent. After some initial success, the Stukas soon suffered heavy losses and were withdrawn from the battle. The bombing of the RAF airfields was continued using Heinkel HE-111, Dornier DO-17, and Junkers JU-88 twin-engined bombers in an effort to decimate

smoke. "Laurel, that's the school receptionist, she told you that nobody gets in to see the Professor without an-"

"Now wait a minute," yelled Blakeley. "I'm not going through that again-"

"Settle down, sugar," said Hilda. "I'm just trying to tell you. Not Laurel, not Leo Junior, not me, not Hoppity over in the diner, nobody gets in to see the Professor. He's a very private man. He's got that thing in his head, you know. That . . . dent."

Blakeley threw his arms up in exasperation.

"Why can't I get anybody around here to understand that there's a war on?" He yelled.

"Keep it down, sugar," said Hilda. "I've got night workers. What I'm trying to tell you is, if you want to see the Professor, just come to the Club tonight."

Blakeley looked puzzled. "What do you mean?"

"It's his only recreation," said Hilda. "He's in the Club here every night, like clockwork. You can't miss him. Any of the girls can point him out for you. Besides, Leo is always right there with

The Junkers Ju-87 was a superior tactical weapon for the Germans in the initial stages of the war. However, Stuka squadrons suffered heavy losses at the hands of the RAF when employed for strategic bombing in the Battle of Britain. (Above photograph courtesy of the U.S. Navy. Inset photograph courtesy of the Royal Air Force Battle of Britain campaign diaries)

Firemen battle a blaze in a heavily bombed section of London during the Blitz, 1941. The terror bombing of civilians ultimately backfired on the Germans, as it strengthened the resolve of the British people while simultaneously providing a critical relief from the bombing of the RAF airfields. (Photograph courtesy New York Times Paris Bureau Collection, National Archives & Records Administration)

the RAF fighter forces.

In this manner, Germany segued into a strategic bombing campaign against the British Isles. In mid to late August, inclement weather conditions and target confusion caused the inevitable bombing of civilian centers of population, with the inevitable casualties. Britain retaliated by bombing German population centers, and the Germans escalated their bombing of the cities. Ironically, the bombing of the civilian populations centers gave the RAF a crucial respite, allowing them to regroup and ultimately repel the German aerial invasion.

Thus was the pattern of aerial warfare cast. The remainder of the war would see an escalation of strategic bombing attacks on fortress Europe by the Allied Air Forces. The German designers would develop ever more exotic aircraft to defend the Fatherland against the Allied bombers. The Allies would develop more powerful and long range fighters to defend their bombers. It was into this heady race for air superiority that the gauntlet of Professor Hermann Flungk's brainchild, the Z-44 Mark I "Boomerang," would soon be thrown.

him, sucking up to the Professor and putting his drinks on the Professor's tab. You'd recognize Leo, even with that fake mustache he draws on his lip every night."

"How unusual," said Blakeley. "A straight answer. Thank you, Hilda. Or, I suppose I should say, Miss Kitty Kat."

Hilda blew another smoke ring.

"Just doing my part for the war effort," she said.

Postcard of Count von Zeppelin's Airship 4. Designed for peacetime passenger service, the Zeppelin soon became an instrument of war. (Bain Collection, Library of Congress)

95

11. MEOW!

Above: The logo of Miss Kitty Kat's Top Hat Club, as seen on the famous cocktail napkin containing the first concept drawing of the Flungk Z-44 "Boomerang." (Image courtesy of the Whitley Speale Collection of the Bone Lake Research Museum)

(Excerpt from *The Flight of the Boomerang*, by Elmer C. Wackmallit, continued.)

When Blakeley awoke, it was dark. He fumbled for the pull switch on the bedside table lamp, and looked at his watch. It was a quarter past ten.

Blakeley was thinking that he must have been more worn out than he thought. Of course, there had been no shortage of stress in his life of late. He went into the bathroom and groped in the dark for the pull switch on the lamp over the mirror. He ran some cold water over a washcloth and wiped his face with it.

Blakeley walked back into the bedroom and sat on the corner of the bed. His stomach was growling ferociously as he pulled on his brown oxfords, but he was far more concerned with going to the Club and finding the Professor.

As Blakeley closed the door, and was turning the key in the lock, the door across the hall opened. Into the hall way stepped Della, the girl with the

Della Bergenstraum at her desk in the offices of the Z.S.ofA.M.&B.T. The door to Professor Flungk's inner sanctum is in the background. The shadowy out-of-focus figure by the door has never been positively identified, although some authorities, most notably Dr. Waldo von Heinkerblonker himself, have claimed that it is Harvey Peastone. (Photograph courtesy of Halloway Bumpsteed, Jr.)

Betty Grable hair.

"Soldier boy!" she said brightly. "I didn't know we were neighbors!"

She was no longer dressed in the form fitting dress he had seen her in earlier. Now she wore an equally form fitting pink fuzzy bathing suit, black fishnet stockings, and red strap heels. In her hand she carried a red clutch purse and a black half mask with cat ears. She blew a pink bubble, and, popping it, expertly rounded the blown bubble off of her lips with her tongue.

"Yes, I suppose we are," stammered Blakeley, somewhat at a loss for words. "I'm sorry, I didn't get your name earlier, Miss . . ."

"Della Bergenstraum!" she said. "But none of that 'Miss' stuff! Call me Della. Everybody else does! What do they call you?"

"Is that your usual evening attire?" Blakeley asked, a little stiffly.

"Is it just me, or did it suddenly get stuffy in here?" Della frowned.

"I'm sorry," Blakeley sputtered. "I suppose I am unaccustomed to seeing . . . well, so much of a young lady. My name is Julian, by the way. I didn't mean to be rude."

There was a clatter of spike heels from within the room. Laurel, the receptionist, pushed out the door behind Della. Suddenly, the little hallway was very crowded.

"Della, do you have your key?" she asked. "We'd better hurry. Saturday is our big night, and you know how Hilda gets when we're late for inspection! She could give Mussolini dictator lessons!"

Laurel was dressed in an outfit similar to Della's, only hers was purple. Della snapped open her purse and handed the room key to Laurel. Laurel locked the door and turned. It was only then that she saw Blakeley.

"Hmph!" she snorted.

"Ah, Miss Congeniality!" said Blakeley. "We meet again!"

"Laurel's right, Julie," said Della. "May I call you Julie? We have to get to the Club. Maybe we'll see you there later, huh?" She expertly popped another bubble.

"As a matter of fact, that's where I was headed just now," said Blakeley.

"Won't be open for another hour," said Della. "You might as well go over to the Diner and put on the feedbag. Tell Hoppity that Della sent you. After all, a growing boy like you has got to eat."

"That's a good idea!" said Blakeley. "I will at that. I could eat a horse!"

"You probably will," said Laurel. As both girls turned and hurried down the narrow hallway, Blakeley watched their cat tails wagging behind them.

"Wait!" Blakeley yelled after them. "How did you know I hadn't already eaten?"

"Nobody ever told you that you snore like a freight train?" Della turned her head, smiled and winked at Blakeley.

"See you later, Soldier boy," she said.

MISS KITTY KAT'S TOP HAT CLUB

After a plate of chicken-fried steak and fries, Blakeley walked back through the bristle of wild shrubbery and crossed the Motor Inn parking lot. In the short time that he had been in the diner, the parking lot of the Blue Moon Motor Inn had filled up.

Cars continued to stream down the highway from town, some parking on the side of the highway. Blakeley joined the crowd of people flowing toward the Motel. At the front entrance to the Motel there was a noisy crowd of people who were funneling through the door. Blakeley joined the group, and, once inside, moved with the boisterous stream. Closer to the entrance, there was a table where Hilda sat, still in her housecoat, although she had removed her yellow shower cap, and replaced it with a roaring twenties cloche hat, complete with a bountiful array of ostrich feathers sticking up from a huge rhinestone medallion on the front. She was methodically taking

Laurel Lautermilk is pictured here in a slightly different attire from her usual office apparel. Although the date of this photograph is uncertain, it could very well have been taken on the occasion of the first flight of the Flungk Z-44. (Photograph courtesy of Halloway Bumpsteed, Jr.)

dollar bills from the slow moving crowd and stuffing them in a King Edward cigar box on the table.

"Hey, sugar," Hilda said to Blakeley as he drew up to the table. "Stag tonight?"

"Just me," he said. "I seem to have alienated everyone else around here."

"Two dollars," she said, and took his bills.

Once inside, it took a few minutes for Blakeley's eyes to adjust to the light, or rather, the lack thereof. The pressure of the crowd pushed him into the room. He saw an empty table along the back wall, and took a seat. As his eyes adjusted, he reconnoitered the room. A five-piece combo had set up on a low stage in the front. They were dressed in tails and top hats, but even from his table in the back, Blakeley could see that they were ragged and threadbare.

The leader, a la Benny Goodman, was vigorously conducting with his clarinet. The band was belting out a thin and somewhat tinny rendition of "Whatcha Know Joe?", although the Gene Krupa drum solo promised to be even thinner, as the drummer had only a snare and a top-hat cymbal. In front of the bandstand was a small dance floor. A few couples were working out variations of the Lindy Hop. The place was packed, and the crowd noise on top of the music made a cacophonous din within the enclosed room.

Blakeley noticed Laurel across the room, carrying a trayful of martinis. At least he assumed it was Laurel; he couldn't see her face as she was wearing her black Kitty Kat mask. But she was the only one with a purple costume. Other girls circulated through the crowd, in their skimpy Kitty Kat costumes, carrying drink trays. Blakeley saw a bright red costume, a deep midnight blue, and a bright green, but no one else he recognized.

"What's your poison, soldier boy?" Della placed her drink tray on his table and sat in the chair beside Blakeley.

"What?" Blakeley cupped his hand behind his ear. "I can't hear you!"

Della frowned, and slid a martini off her tray, placing it in front of Blakeley. "That one's on the house," she shouted in his ear. "I'll just tell Hoppity that the Professor knocked it over."

"Hoppity works the bar here too?" Blakeley asked.

Professor Flungk in one of his more light-hearted moments at Miss KittyKat's Top Hat Club. Normally reticent to the point of being a virtual recluse, the Professor's lighter side always came forth with a liberal application of peppermint schnapps. (Photograph courtesy of the Whitley Speale Collection of the Bone Lake Research Museum)

The fate of the entire nation may hang in the balance! That piece of junk has been gathering dust in the back of the hangar for years, and nobody gave it a second thought until you came along. What's the big deal?"

Blakeley opened his mouth to speak, but Della interrupted him before he got started.

"Aw, I don't care anyway," she said. "He has a reserved table down front. You see Mandy, the cigarette girl?"

Blakeley looked to where Della was pointing. Walking down the center aisle, working her way in the direction of the bandstand, was a brunette in a dark green bell-hop costume, complete with brass buttons and red piping. A tiny pillbox hat was perched at a jaunty angle on her head. Instead of pants, she wore a short short matching pleated skirt. She carried a cigarette tray suspended by a ribbon around her neck, and wore a black garter belt on her left thigh.

"The Professor's table is right in front of Mandy, next to the dance floor. "You'll probably see Leo jump up any minute to grab a cat tail," said Della. "That juvenile delinquent. Well, I've got work to do. Maybe I'll see you later."

Della picked up her drink tray and melted into the crowd.

Blakeley stood up to get a better look. He saw Laurel in her purple Kitty Kat suit, working her way around the perimeter of the dance floor. Sure enough, when she passed the table where Della told him the Professor was sitting, he saw a character in a white sports coat, epaulets and what looked like a taxi driver's hat, jump up and grab her tail when she went by. Without spilling a drink on her tray, Laurel expertly turned, slapping him across the side of his head. The taxi hat flew onto the table. Without his hat, Blakeley immediately recognized Leo. Laurel picked up her tail, and walked haughtily away. Laughter erupted from the nearby tables.

Martini in hand, Blakeley worked his way through the crowd. As he neared the Professor's

"Everybody does double duty around here. Don't you know there's a war on?" she said, winking at him. Della got up from the table. "Well, a woman's work is never done. Catch you later!"

"Wait!" said Blakeley, grabbing her arm. "The Professor. Where can I find him?"

She looked down at her arm "Lookee, no touchee!" she said, and peeled his hand gently off her arm. "House rules!"

"Sorry," Blakeley stammered. "But I really must find the Professor. It's-"

"I know, I know," she said. "It's of vital importance to the defense of the good old U.S.A.!

table, he had closed to just a couple of feet behind the cigarette girl. Ahead, he could hear a high pitched squeaky voice, yelling "Cigaretta! Cigaretta!" with a Germanic accent, followed by maniacal laughter.

As the cigarette girl approached the table, a hand reached out from the table and deftly pulled her garter. It rebounded with a loud 'Snap!'

The girl yelled "Ouch!"

Maniacal laughter erupted again from the table, and the girl said "That's really cute, Professor!"

As Blakeley approached the table, the hand that snapped the garter now held a cigarette, European style, between the thumb and forefinger, with the hand turned outward. Blakeley came around the cigarette girl just as she bent down to light the cigarette.

In the glow of the match, Blakeley could see a diminutive man with sparse reddish brown hair, in a neat pinstripe suit and conservative tie, wearing round horn rimmed glasses. His most distinguishing feature,

Extremely rare group shot of some of the employees of Miss KittyKat's Top Hat Club. Featured are Laurel Lautermilk, standing; in the second row from left to right, Amanda Gunderson (nee Smilch) in glasses, Della Bergenstraum, and Esmeralda Grackle. The others are unknown. This photograph was likely taken on the occasion of one of Professor Flungk's outdoor summer festivities, even though Whimpington maintains that it was taken on the day of the first flight of the Flungk Z-44 Boomerang. ("Whimpington fabricates without the slightest shred of evidence!" Bumpsteed) Irrespective of the ongoing controversy as to the date, the fact remains certain that the girls are celebrating the occasion with Coca Colas from the very machine which would later prove to be the undoing of the unfortunate Buchard Woolsey. (Photograph courtesy of Halloway Bumpsteed, Jr.)

however, was a deep indentation across his forehead.

LT. BLAKELEY MEETS THE PROFESSOR

"Excuse me," Blakeley addressed the man. "Are you Professor Flungk, the designer of the partially destroyed aircraft in the Technical School Hangar?"

"Commodore!" exclaimed Leo, jumping up and grabbing Blakeley's hand, shaking it vigorously. "Fancy meeting you in here, of all places! How are those tires workin' out for ya?"

The Professor looked up at Blakeley uncomprehendingly, then at Leo. "Vas is los?" he asked.

"Commodore!" said Leo, ignoring the Professor. "Pull up a chair! We was just talking about ya!"

Leo pulled out a chair, inviting Blakeley to sit. "Professor, this is the moke I was just telling ya about!"

"Ya?" asked the Professor.

"Professor, I'm Lieutenant Julian Blakeley of the U. S. War Department Office of Special Projects," Blakeley rattled off enthusiastically. "My department is very interested in the aircraft that you developed a few years ago-"

"Whoa! Slow down, Commodore!" Leo

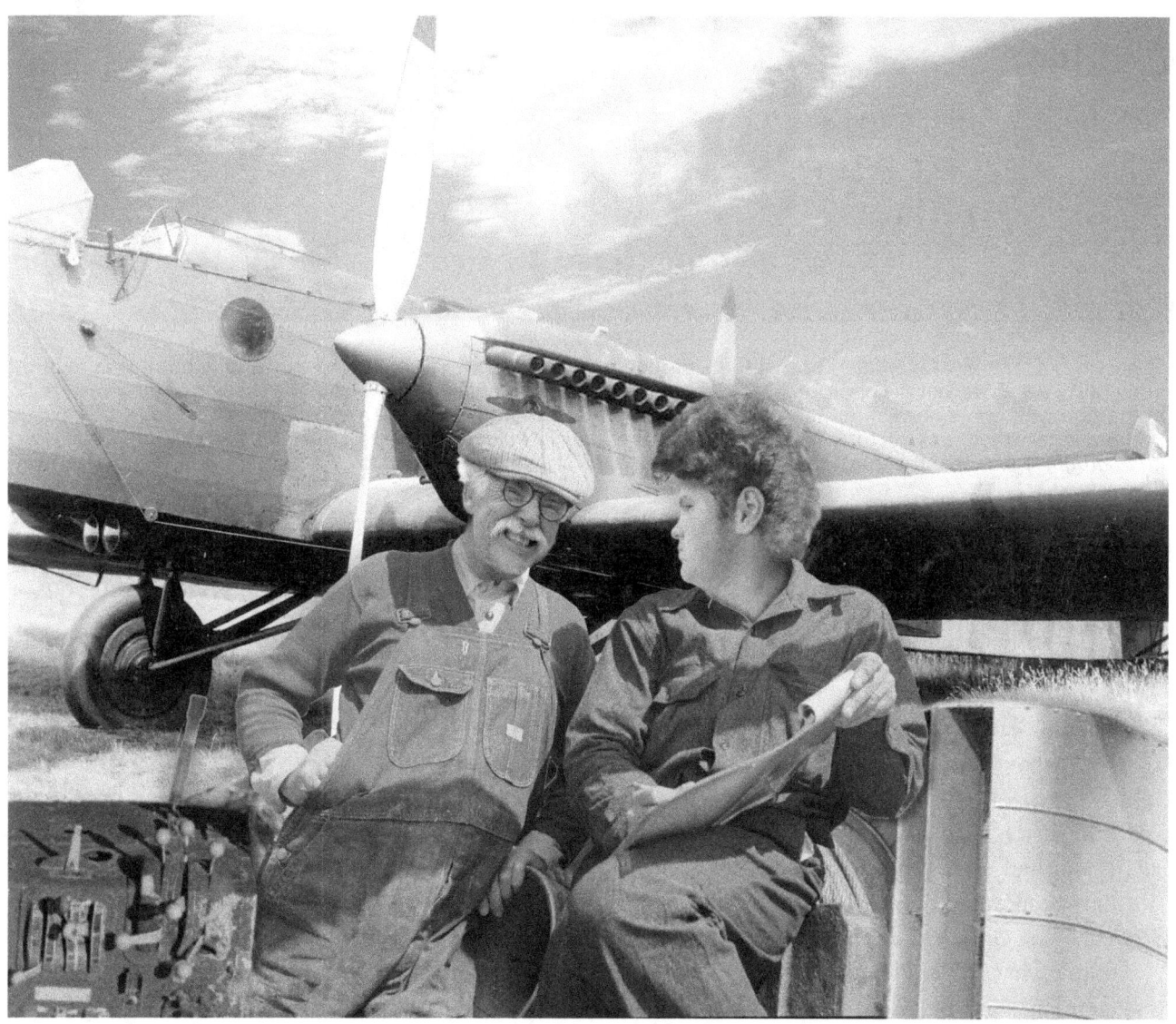

Harvey Peastone, left, and Leo Hooper, Junior, right, confer on a technical matter during the latter part of the design phase of the Flungk Z-44 "Boomerang," which is shown in the background. Although he rarely, if ever, performed any actual work on the Z-44, Peastone liked to wear his coveralls around the facility and just be "one of the boys." (Photograph from the private collection of Dr. Eustis Bothomfieder.)

Street entrance to Miss KittyKat's Top Hat Club, probably taken in mid 1941; the date is uncertain, but it is agreed by nearly all researchers that the photograph was taken before the club burned down. Note the spelling of the word "club" in the sign. This offers nearly irrefutable evidence of the spelling of the club's name, and should put to rest Dr. Eustis Bothomfieder's theory, as put forth in his epic treatise *The Veil of Stupidity: Code Level Ultra-Z Programs, Scams, Cover-Ups, and Outright Frauds*, that the true name of the club was "Miss KittyKat's Top Hat Klub." (Photograph courtesy of Halloway Bumpsteed, Jr.)

interrupted. "The Professor, he don't speak regular like you and me. All he knows is Kraut talk. I'm sort of his interloper."

"You mean, his interpreter?" asked Blakeley.

"Schnapps?" asked the Professor, holding out a shot glass to Blakeley.

"Yeah, that too," said Leo. "His - what did you call it?"

"Interpreter," said Blakeley.

"Yeah, that!" said Leo.

"I'm somewhat surprised that you speak German," said Blakeley.

"Speak what?" asked Leo.

"German," said Blakeley. "Kraut talk."

"Oh, yeah, well, I don't actually speak it," said Leo. "I just listen to what you have to say and then I sort of just tell it to the Professor here, until he gets it. It ain't easy. Sometimes, it takes a while, let me tell ya!"

"Pretzel?" asked the Professor, holding out a bowl to Blakeley.

"Thank you," said Blakeley, grabbing a handful of Pretzels. "Uh, danke."

"Ahh!" explained the Professor. "Sprechen sie Deutches?"

"No," said Blakeley. "Uhh, nein, mein herr. Only English."

"Like I was saying," said Leo, trying to reestablish his momentum. "Anything you want to say to the Professor here, you can say to me. I'll see that he gets it. The Professor and me go back a long way, don't we?"

"Ya, ya," said the Professor, emptying his shot glass of Schnapps.

"Did you say that you were just discussing my situation before I arrived?" asked Blakeley.

Enrollment ad run by Zanesville School of Aviation Maintenance & Beauty Technology in the late thirties. This ad appeared primarily in comic books and Hollywood gossip magazines and was an effective recruitment tool for the school. However, as Bumpsteed claims, the ad considerably oversold the charm factor of Z.S.ofA.M.&B.T., particularly with the inclusion of palm trees, which at the time were not normally found on southeastern Ohio airports. (Artifact courtesy of the Whitley Speale Collection of the Bone Lake Research Museum)

"Don't that beat all!" said Leo. "I was just telling the Professor here, that with your tire purchase today, that you qualify as a Zanesville School of Aviation Maintenance & Beauty Technology preferred customer!"

Laurel walked by, and the Professor grabbed her tail and yanked it. She immediately swung around and smacked Leo on the head, knocking his hat onto the table.

"Mein Kitty Kat! Schnapps! Schnapps! Bitte, Liebling!" The Professor cackled, banging his Schnapps glass on the table.

"Hey, what did ya hit me for?" Leo yelled at Laurel. "I didn't do nothin!"

"I know," said Laurel sweetly, as she picked up a cut glass bottle from her tray and filled the Professor's shot glass. "But if I hit the Professor, I'd lose my job. You wouldn't want that, would you?"

"Don't bet on it," said Leo sulkily.

"Yeah," Laurel said. "Remember that the next time you want me to cover for you when you're too hung over to make it to class."

"That was the flu!" said Leo, petulantly.

"I see you finally got your appointment with the Professor," she said, turning to Blakeley.

"Danke!" said the Professor. He held up his shot

This highly suspect photograph, purportedly of two Z.S.ofA.M.&B.T. students salvaging parts from the wrecked Z-44 "Boomerang," has repeatedly been foisted off on the American public as an authentic representation of the state of the aircraft following Hooper's ill-fated flight. No less august authorities than Dr. E.C. Whimpington, and the illustrious Dr. Eustis Bothomfieder have used this photograph in their texts to claim that the damage to the Z-44 was much more severe than claimed by Blakeley. As Halloway Bumpsteed, Jr. states in his massive volume, *They Might Have Been Giants: Misadventures, Blunders, & Colossal Failures in Aviation*: "Whimpington doesn't know a propellor from a pumpernickel, and Bothomfieder writes as though he does his historical research on the backs of cereal boxes." (Photograph courtesy of the Whitley Speale Collection of the Bone Lake Research Museum)

glass in a toast to the table, before slugging it back.

"Yeah!" said Leo. "I got it all worked out for the Commodore with the Professor here!"

"Great things will come of this," she said. She refilled the Professor's glass, turned and walked off.

"Hey!" yelled Leo. "What about my drink?"

"Pull somebody else's tail," she said over her shoulder, before disappearing into the crowd.

"Whoa!" Leo exclaimed.

"Indeed!" said Blakeley.

"Ya, ya!" said the Professor.

This photograph, taken in the winter of 1943, plainly puts the lie to Whimpington and Bothomfieder's dubious claim that the Flungk Z-44 "Boomerang" was totally destroyed following Hooper's flight. It shows the ill-starred craft in its final resting place, in a war-drive scrap heap on the outskirts of Zanesville, Ohio. However, the Z-44 would soon rise Phoenix-like from its own ashes, as its valuable components would shortly be melted down and reforged into weapons to protect the American people from the shackles of tyranny. (Photograph courtesy of Halloway Bumpsteed, Jr.)

THE NEGOTIATION

"So, did you mention to the Professor that I was very interested in his aircraft?" Blakeley asked.

"Well, Commodore, I just said we were discussing your situation-"

"With me being a preferred customer now," said Blakeley.

"Yeah!" Leo said. "You know that automatically gives you a ten percent discount on all of our products and services."

"Too bad I didn't have that before I bought my tires," Blakeley said.

"Yeah, that's too bad," Leo said. "Say, you don't have any motorcycles, do you?"

"Motorcycles?" Blakeley asked.

"Yeah, it doesn't matter even if they're green, like your car," said Leo. "We've got lots of paint."

"Why would I have a motorcycle?" asked Blakeley.

"Don't the navy have some motorcycles, you know, in that-what did you call it-pool hall, or swimming pool, or something like that?" asked Leo.

"You mean the motor pool?" asked Blakeley.

"Yeah, that's it," said Leo. "Motor pool! I figured if you had one, we could do some trading on that "Boomerang" in the hangar. Beats me, but the girls like motorcycles almost as much as they like a uniform. Heh, heh! Women! Just try to figure one out!"

"I'm really interested in the design concept of the-" Blakeley said.

"Well, you're in luck," Leo interrupted him, "because-guess what?" The corners of Leo's mouth moved upward in his automatic smile.

"At this point, I really haven't the foggiest idea," Blakeley said.

"Don't you get it?" Leo punched him in the arm. "Designing is a service, and you still get your ten per cent discount! Ain't that great?"

"Ein anderes Schnapps, bitte, Liebling!" The Professor was waving his shot glass in the air, trying to get the attention of the Kitty Kat in midnight blue, who didn't seem interested in coming anywhere near the table.

"It's really a little early to discuss financial arrangements," said Blakeley.

The midnight blue Kitty Kat finally came over to their table and filled the Professor's shot glass. Blakeley noticed that every one of the Kitty Kat girls carried a little cut glass bottle of Schnapps on their tray.

"Danke, danke," said the Professor.

"What I would really like to do is talk to the Professor about his aircraft, and perhaps look at his plans." Blakeley said.

"Ya, ya!" said the Professor, smiling widely. He raised his glass in a toast to Blakeley. "Aero plans!"

"Yes," said Blakeley, smiling back at the Professor and raising his martini glass in return. "Aero plans!"

"Sind sie ein Flieger?" asked the Professor.

"What did he say?" Blakeley asked Leo.

"He wants to know if you got any motorcycles to trade," Leo said.

A gal of many talents, one of Della Bergenstraum's many duties at Z.S.ofA.M.&B.T. was to maintain the proper fluid levels on Professor Flungk's 1939 Packard, even though he rarely left the school environs. (Photograph courtesy of the Whitley Speale Collection of the Bone Lake Research Museum)

DON'T SHOOT ME, I'M ONLY THE PIANO PLAYER

Blakeley stared at Leo in utter disbelief.

"Hey! Didn't I try to tell ya a minute ago?" Leo said. "The Professor may be crazy, but he knows what he wants!"

"I don't know any German, but I don't think 'Flieger' means motorcycle." Blakeley said.

"Ya!" the Professor said, putting his flat hands together and making flapping motions with them. "Ein Flieger! Sind sie ein Flieger?"

"Flieger is Kraut talk for Indian," said Leo. "He must want an Indian motorcycle. I wouldn't be so picky myself, I'd be happy with a Harley-Davidson."

"The Professor started flying his hands around in front of himself. He was making motor noises with his mouth, spraying Blakeley with Schnapps in the process."

"He's talking about an airplane!" said Blakeley.

"Ya!" said the Professor, smiling. "Ya! Aero plans!"

"If we got a Harley-Davidson," said Leo, staring at the ceiling in deep thought, "I'll bet you could find a couple of Indian emblems and stick them on the tank. Nobody would ever know the difference."

Blakeley pointed at the Professor. "You have an aircraft, in the hangar across the street . . ."

The Professor looked puzzled.

"Nicht ben ein Flieger!" he said.

"No, no," said Blakeley, pointing at the Professor again. "Your 'aeroplans!'

"He wants to know about your boomerang," said Leo.

"Ya! Ya! Das boomerang!" said the Professor, his face brightening.

"Ya!" said Blakeley, excited now. "Das Boomerang! Vroom! Vroom!"

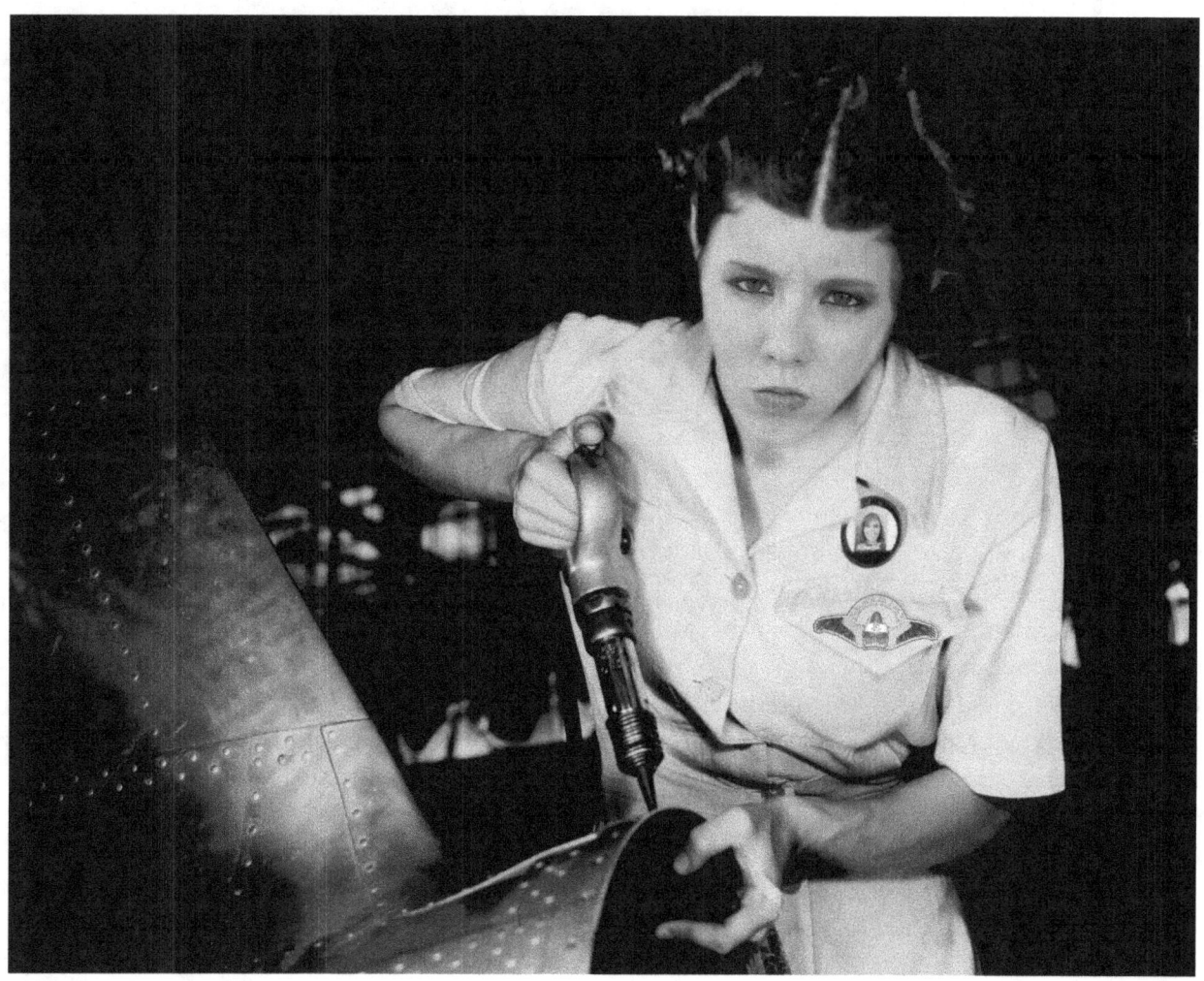

Although Laurel Lautermilk was occasionally called upon to work in the shop like all of the other girls, she was never too happy about it. (Photo courtesy of Dr. E.C. Whimpington)

"Ya!" said the Professor, flying his hands once more. "Vroom! Vroom!"

Then his hands took a nosedive into the table. "Und kaput!"

"I still think he wants a motorcycle," said Leo.

"You keep out of this!" said Blakeley.

"Whaddya talking about?" said Leo. "Hey, you and me, we had a deal, remember? I said I'd get yas in to see the Professor, and you'd get me a motorcycle!"

"I said nothing of the kind!" Blakeley said. "Maybe the best thing for you to do is take a hike, and let me talk to the Professor myself!"

"Commodore, you got some nerve!" Leo said. "You're making a big mistake! I'm the Professor's right hand man. He doesn't make a move without my say-so! Ain't that right, Professor?"

"Ein anderes Schnapps, bitte," the Professor said.

"See!" Said Leo. "What did I tell ya?"

Blakeley saw Laurel a couple of tables away, caught her eye and said, "Another Schnapps for the Professor, Liebling."

Laurel came over to their table, poured the Professor a fresh glass of Schnapps, and, to Blakeley's surprise, began to converse with the Professor in fluent German. They talked animatedly for three or four minutes, then Laurel turned and started back up the aisle.

"Wait!" Blakeley jumped up from the table and grabbed Laurel by the shoulder. She looked down contemptuously at his hand, which he quickly withdrew.

"Sorry, I forgot," said Blakeley, air dusting her shoulder. "Lookee, no touchee. But- you were talking to the Professor-"

"Of course,' she said demurely. "I'm really not the iceberg you seem to think I am."

The irascible and ever cantankerous Halloway Bumpsteed, Jr., perpetual thorn in the side of Bothomfieder, von Heinkerblonker, Whimpington, et al. Bumpsteed's opinionated rants on the subject of the Flungk Z-44 "Boomerang" forever earned him the enmity of aviation historians everywhere. (Photograph courtesy of the Whitley Speale Collection of the Bone Lake Research Museum)

"No!" he said excitedly. "It's not that! I mean, I don't think you're an iceberg! But, you speak German! The Professor's language!"

"Of course," said Laurel. "German is my first language."

"That ain't German!" Leo said. "That's gobbledy gook!"

"I don't understand," Blakeley said, ignoring Leo.

"I was raised Amish," Laurel said. "German was

the only language I spoke growing up. Why do you think I got the job as receptionist here?"

"That's just hogwash she's talking!" said Leo. "The Professor's just playing along with her because she's a girl!"

"Ya," said the Professor brightly, pointing to Laurel. "Mein kliene Liebling!"

"I thought I told you to keep out of this!" Blakeley glared at Leo. Then, to Laurel, he said, "You're not exactly dressed Amish."

"An Amish girl has one career choice. Marry some lovable lunk with a beard, raise eight kids, and work on the farm all day," said Laurel. "I wanted some other options, that's all."

"What were you and the Professor talking about?" Blakeley asked.

"He asked me who you were and what in the world you were talking about," said Laurel. "I told him you were from the U.S. government and that Uncle Sam was very interested in his Boomerang."

"Really?" said Blakeley, grinning like a kid in a candy store. "What did he say?"

"Hey, wait just a minute!" yelled Leo, jumping up from his chair. "The Professor can't do a thing without me! You think you'd get a minute of work out of those lug nuts over at the school without me standing over them? And who do you think flew the Boomerang? Me! That's who! You two got some nerve, that's all I got to say!"

"Ein anderes Schnapps, bitte, Liebling," said the Professor.

Laurel smiled at the Professor, and, walking back to the table, poured him another shot of Schnapps. Ignoring Leo, she addressed Blakeley, "He said to stop by tomorrow, and he'll tell you all about the Boomerang. Of course, I'll be there to translate for you."

"You think you can cut me out just like that, you little Amish tart!" Leo reached into the inside pocket of his sports coat and pulled out a vicious looking switch-blade. He flipped the switchblade open, his trademark maniacal grin spread across his face. Laurel stumbled back, real fear showing in her eyes. Leo pushed back his chair, and it fell, clattering to the floor. He stalked around the table toward Laurel.

Blakeley reached under his coat and pulled the flare pistol from his waistband. Stepping between Leo and Laurel, he leveled the pistol at Leo.

"That's not very smart, Leo," Blakeley said evenly. "Bringing a knife to a flare gun fight."

Suddenly the band stopped playing. The room grew deathly quiet. Leo stared down the stubby barrel of the flare gun. It looked as big as a cannon. He stopped in mid-stride. He looked down at his hand holding the knife as if it was a thing alien to him.

"Whoa!" he sputtered, backing up. He dropped the knife, and it clattered loudly, resounding around the deathly quiet room. "I was just kidding! Heh! Heh! It was a joke, right? Had you guys going for a minute, there didn't I?"

Leo continued backing away from the flare gun. "Please, Commodore!" Leo said. "Don't shoot me, I'm beggin' yas!"

Then he tripped backward over his fallen chair. His feet flew up and knocked over the pedestal table. The Professor's Schnapps went flying at Blakeley as the edge of the table smacked him squarely in the mid-section. Blakeley tumbled backward with a loud "Whoof!" of exhaling breath. Blakeley hit the concrete floor with an audible thump, reflexively squeezing the trigger of the flare pistol. There was a loud "Crack!" from the pistol and the flare shot to the ceiling, bounced off, and arced across the dance floor, coming to rest with a loud crash behind the band stand.

The band immediately started playing a screechy rendition of "We're in the Money!", like a phonograph record playing at high speed. A few couples on the dance floor actually started dancing again, until flames erupted in the curtains behind the band.

Someone yelled "Fire!" and the crowd began a mad dash to the exits. To the band's credit, they continued to play until the heat of the fire chased them off the stage.

Luckily, someone had the presence of mind to call out the airport fire truck. It was three AM before the last glowing embers of what was once Miss Kitty Kat's Top Hat Club were extinguished.

AFTER THE PARTY

"I what?!?" Blakeley tried to sit up in bed, but immediately his head pounded like it was inside a cathedral bell.

"The girls are a little upset with you," said Hilda, laying a cold washcloth across his forehead. "Lucky for you we had fire insurance, or I would have gutted you like a catfish while you were still unconscious."

"Ooohhh!" moaned Blakeley. "Where am I?"

"You're in your motel room," said Hilda, taking a drag on her unfiltered Pall Mall and blowing a smoke ring. "It's a good thing you only burned down the club, or we'd all be out in the street."

"The last thing I remember-"

"The drummer wants to talk to you about suing the U.S. government for his drum set. I only recall seeing a snare drum and a cymbal, but now it's a whole double bass rig with tom-toms, just like Gene Krupa's," Hilda said.

"My head!" said Blakeley. "What-"

"Your flare pistol went off," Hilda said. "Nobody was hurt really bad. A couple of broken bones and some bruises in the stampede is all. Even Leo survived, unfortunately. That boy is my poor dead sister Zelda's spawn, but I would have parted his hair with my double barrel twelve gauge if he had hurt one of my girls."

Blakeley tried to sit up again. "I've got to-"

"See the Professor, I know," Hilda said, placing her hand in the middle of his chest and gently pushing him back down in the bed. "Laurel stopped by this morning and said to tell you the Professor's ready to see you as soon as you're up and about. She brought you a take-out carton of chicken soup from the Diner, too. It's on the night stand when you're ready for it."

Hilda Grackle, alias Miss KittyKat, was always willing to pitch in for a publicity photo op, as long as it occurred after two P.M., and she didn't have to actually work. (Photograph courtesy of Halloway Bumpsteed, Jr.)

Hilda got up from the bed and walked to the window. She pulled the curtain closed.

"Right now, you need some rest," she said. "We couldn't get a doctor last night, but Hoppity looked at your eyes and said that you had a mild concussion from cracking your head on the concrete floor. But don't worry, Leo had a worse concussion than yours, too. He's still out cold. Anyway, thanks for standing up for Laurel last night. You do remember that, don't you?"

"Vaguely," said Blakeley.

"That's another reason that I didn't gut you like a catfish," Hilda said. "Now, you get some rest."

Hilda walked out of the room, quietly closing the door.

In the darkness, Blakeley fell into a deep and restful sleep.

THE LIEUTENANT'S PARABLE

How fortuitous the fates, says the god of hindsight,
That Blakeley met Flungk, on that glorious night.
Where Leo, faltering, drew forth a knife,
Then thrust at poor Laurel, to cut short her life.
And gallant Julian, midst all the hubbub,
Shot off his flare pistol in the Kitty Kat Club.
Thus in so doing, he vented his ire,
Though none were injured in the ensuing fire.
How fleeting the honor, how empty the fame,
Of he who did venture in defense of a dame.
For who now remembers the angst and the drama,
Of the night the Lieutenant put Leo in a coma?
And who now recalls, though it started with a bang,
This humble beginning of the great whirling "Boomerang."

From *Daedalus Unleashed: Poems of Wind and Sky*
by Leslie Collard Hapgood

12. TEMPERED IN THE CAULDRON OF CHANCE: THE EARLY LIFE OF HERMANN FLUNGK

"Boomerang: A peculiar missile used in hunting and war. It takes the form of a curved stick about two feet long. It prescribes peculiar curves according to the shape of the instrument and the manner of throwing it, and taking a backward direction, falls near the place from which it was thrown, or even behind it."

Webster's New 20th Century Dictionary of the English Language

The Teutonic Knights Lodge in simpler times, before the Great War. This photograph was taken on an earlier, happier Hedge Hog Day, February 2, 1911. Professor Flungk's parents are seated on the front row: Hermann, Senior at far right, and Helga, second from right.

In order to better understand the events of the prevous chapters, and subsequently, the technological developments attributed to Professor Flungk, an examination of the formative years of the Professor is indicated. Truly relevant information, unfortunately, is almost non-existent. Nonetheless, the author has diligently searched for all possible facts related to the Professor's psychological maturation. Those results are presented here.

To this day, little is known of Prof. Flungk's early life. The sole relevant fact of aeronautical significance, uncovered only after years of tedious and mind-numbing research, is that he was known to be obsessed with boomerangs from an early age. This obsession came about through a series of seemingly random events, which nonetheless set the youth upon a course which, in retrospect, would appear to lead inevitably to the development of the Professor's life work. While he was still an infant, young Herrmann's parents, recent immigrants to the United States from Austria, were expelled as sympathizers to the German and Austro-Hungarian cause in 1917. Although there is no direct evidence that the elder Herr Flungk had engaged in any pro-German activities, court and incarceration records found in the Allegheny County courthouse indicate that he was arrested in the early morning hours of February 3, 1917, with a group of thirty-seven cohorts, at the Teutonic Knights Lodge in Frankfurter, Pennsylvania, where they were engaged in an extended celebration of "Mariä Lichtmass," or "Hedgehog Day." The elder Flungk was charged with "public intoxication and disturbance of the general civil order." This evidently occurred after the eighty-eighth rendition of "Wer, wer, wer ist mein Kleiner Hübscher Kuchen?" at three A.M., which finally prompted irate neighbors to call the appropriate authorities.

Herr Flungk was ordered deported back to his native Austria. Unwilling to leave his wife and son behind, the entire Flungk family was processed out of the country. A bored and possibly semi-literate immigration clerk misspelled "Austria" as "Australia" on their deportation papers. Thus were the Flungks inadvertently booked for passage on a tramp steamer which, in the course of a year-long journey, included one brief stopover at Matilda's landing, a remote mission outpost in North-western Australia. Here they were summarily dumped, their passage having been paid no further. As the Flungks spoke only German, it was four years before they were able to communicate their plight to the proper authorities. By then, the Great War was over, and the elder Flungk was allowed to bring his family back to the United States.

THE "AUSTRALIAN" YEARS: MOLDING A MALLEABLE MIND

This "Australian" period constitutes the most mysterious phase of Prof. Flungk's life. Dedicated

scholars of the ZXPVT-1 "Flungk" Boomerang, as well as other aviation historians, have long debated what transpired during this time to mold (some use the term "warp") the mind of young Herrmann in those most formative years.

Dr. Waldo von Heinkerblonker, M.S., Ph.D., L.S.M.F.T., a staunch defender of the work of Prof. Herrmann Flungk, has extensively studied the early life of Prof. Flungk, with special emphasis on his years in the remote outback of Northwestern Australia. In fact, Prof. Flungk's case was exhaustively theorized upon in Dr. von Heinkerblonker's seminal work,

One can only imagine young Hermann Flungk's distress: a stranger in a new land, unable to communicate with the local tribesmen in their native tongue, and subjected to repeated boomerang attacks by the aboriginal tribesmen. (Photograph courtesy of Halloway Bumpsteed, Jr.).

"Whackos, Crackos, Sickos and Psychos--A Psychological Study of Fruitcakes and the Nutcases who Warped Them."

Although no official records exist of the Flungk family's Australian ordeal, Dr. von Heinkerblonker theorizes that the young Herrmann, (he was then only four), was captured by wild Aborigines. How

Students of the history of the Flungk "Boomerang" owe a debt of gratitude to the inimitable Dr. Waldo von Heinkerblonker, whose epochal work remains the definitive text on the early years of Professor Hermann Flungk. The above photograph was taken just days before Dr. von Heinkerblonker's commitment.

he came to this conclusion, he does not make quite clear. Nonetheless, according to Dr. von Heinkerblonker's theory, young Hermann (he was not yet a Professor) was held captive for some time in the outback, and used on numerous occasions by the aboriginals as a target during boomerang practice.

Wild as this theory sounds, it nonetheless fits the scanty facts. In addition to the thin circumstantial facts in the case, it also neatly explains a large dent in the Professor's forehead, which is strikingly evident in the few surviving photographs that were taken after his Australian sojourn. What is beyond any doubt is that, throughout his career, Prof. Flungk was obsessed with boomerangs. It is no surprise that all of his aeronautical designs reflected this rather peculiar bent.

THE DREAM: FIRST CONCEPTUALIZATION OF THE FLUNGK "BOOMERANG"

Ever since his chance Australian encounter, Professor Flungk's ultimate dream had been to design a whirling aircraft that would possess the unique flight characteristics of the boomerang. His theories predicted that such a craft would be virtually impossible to detect and destroy, due to the high speed spinning of the aircraft. At the same time, the wildly whirling aircraft would undoubtedly send the enemy into a state of confusion. A sub-theorem of his theory, never conclusively proven, was that the boomerang craft, once launched, would always return to base.

An additional point of particular importance to Prof. Flungk, and to no one else, was that the word "Boomerang" was the same in all languages, and would greatly ease problems in translation. This arcane fact reveals a life-long frustration for Prof. Flungk. Having never learned English, he may well have overemphasized the importance of the translative quality of aircraft names.

The early stages of conceptualization of the Flungk Boomerang are shrouded in the hoary mists of history, although some historians claim that these hoary mists were in actuality an alcoholic fog. (See: Bumpsteed, Jr.)

The first design of the Flungk Boomerang is believed to have been drawn up during a planning meeting held by Professor Flungk on New Year's Eve, 1934, although this too is disputed by some historians. (See: Bumpsteed, Jr.)

This design conference purportedly occurred at a local social organization known as "Miss Kitty

This controversial napkin, now on display at the Bone Lake Research Museum in the Whitley Speale collection, is believed by most reputable experts to be the genuine work of the late Professor Hermann Flungk. Of particular interest is the fact that, after many design permutations, the model YPVT-1 as used by the U.S. Army Air Forces during World War II bore a remarkable resemblance to this first draft of Prof. Flungk. (Photograph courtesy of the Bone Lake Research Museum)

The date of the purported design, New Year's Eve, is, according to some historians, significant. (See: von Heinkerblonker, et al) This date would place the conception of the radical new aircraft design coincident with the occurrence of a combination Graduation Party for the Beauty and Aviation School, and New Year's celebration. It has been noted by some of the more critical aviation historians that New Year's Eve was somewhat early to celebrate the school's graduation. (See: Bumpsteed, Jr.)

Indeed, it is true that graduation would have traditionally occurred in May, upon the completion of the school year. However, it should be noted that Z.S. of A.M. & B.T. graduation had also been celebrated on numerous other occasions previously, most notably, Arbor Day, Boxing Day,

Kat's Top Hat Club." Professor Flungk, recognizing the need for a social gathering place for the students, had established the club almost before the ink was dry on the lease papers for the school facilities. Located in the lounge area of the Blue Moon Motor Inn, the facility served the added attraction of providing much needed employment opportunities for the Beauty School Students. Its close proximity to the school and its intimate ties with the institution, therefore, made it a natural location for the Professor to conduct his aeronautical pursuits in a more informal and relaxed atmosphere.

Halloween, and Christmas. It would also continue to be celebrated throughout the coming year and in years to come, on George Washington's birthday, Valentine's day, and Saint Patrick's day, to name just a few.

It is also noteworthy that, according to subsequent inquest records, none of the participants could remember, or claimed not to remember, what had transpired on this particular evening. Prof. Flungk certainly professed no memory of the night in question. It is possible that, due to the secretive nature of the project, this was simply a ruse. Nonetheless, his claim of having no recall of the events in question, may well have been genuine. According to records which the author uncovered at the Sunnydale Rest and Cure Home in Clevisville, Ohio, a "Mr. Herman Flunk" (sic) was interned on January 2, 1934 for "acute alcoholic psychopathy." The records indicate that Mr. "Flunk" was subsequently released on January 16, 1934. Could this Mr. "Flunk" be, in actuality, Professor Hermann Flungk, of aeronautical and cosmetological fame?

Only History knows. Future historians may shed light on the subject in view of additional findings, but for now we must proceed with the facts at hand.

Nonetheless, according to the thesis of Dr. Waldo von Heinkerblonker, at least, the fact that this auspicious date marks the beginning of the Flungk design dynasty is well founded. This specious theory is based solely on the fact that the earliest known conceptual drawing of a boomerang craft, found in the pocket of Prof. Flungk's tuxedo after the night

in question, was undoubtedly drawn on a Miss Kitty Kat's Top Hat Club napkin.

Though many noted aviation historians attempt to cast doubt on the authenticity of this note (See: Bumpsteed, Jr.), or otherwise question its authorship, it is almost certainly attributable to Prof. Flungk, if for no other reason than for the incorporation of a boomerang into the design (reflecting his unfortunate head injury), and the fact that he was a well-known habitué of Miss Kitty Kat's Top Hat Club.

It is interesting to note that the drawing on this napkin most closely resembles the final model of the Flungk Boomerang, even though the craft itself would go through numerous design changes and revisions in its tortured history. This drawing stands in mute testimony to the lone vision of Flungk, a man whose eyesight, if somewhat blurred that night, was locked steadily on a grand, perhaps even grandiose, future. (von Heinkerblonker et al)

13. MAKING THE DREAM COME TRUE:

FINANCING THE Z-44 "BOOMERANG"

Street entrance to the Zanesville School of Aviation Maintenance & Beauty Technology, probably pre-war. The lettered building is the rear of the hangar. (Photo courtesy of Dr. E.C. Whimpington)

Where Prof. Flungk procured the funds for the development of the Boomerang has long been the subject of intense scholarly debate. (See: von Heinkerblonker, Bumpsteed, Jr. et al) However, it is agreed by most authorities that his Aviation Maintenance Assistant Instructor, a Mr. Leo Hooper, Junior, was at least indirectly responsible for the retaining of developmental funding.

It seems that Mr. Hooper, Jr. maintained extensive connections in the alcoholic beverage industry, with which he had been associated before his employment with the Zanesville School of Aviation Maintenance & Beauty Technology.

At Professor Flungk's behest, Mr. Hooper, Jr. evidently contacted Mr. Harvey Peastone, a friend and business acquaintance of Leo Hooper, Jr.'s father, Leonidas "Hoppity" Hooper, Senior. When Leo, Junior was a youth in Chilblain, New Hampshire, Mr. Peastone, who had admired the lad's pluck and native business acumen, had taken Leo, Junior under his wing. Thus, Leo accompanied Mr. Peastone on his many business trips as he delivered his alcoholic product line throughout the northeast. Later, after this internship, Leo, Junior took over the routes himself.

It was the Professor's earnest desire that Mr. Peastone provide some much needed financial assistance to his aviation project, it being common knowledge that Mr. Peastone was a well established New Hampshire businessman with a keen interest in all modes of transportation.

The story of Mr. Leo Hooper, Jr.'s acquaintance with Mr. Peastone is rather convoluted. Nonetheless, that story is presented here, along with additional background material, so that the reader may better understand the events which ultimately led to that unique combination of individuals, materials, and ideas which resulted in what is certainly one of the strangest tales in the annals of aviation history.

Needless to say, these preliminary events were exceedingly strange in themselves. Though the details are scanty, most reputable researchers agree, albeit loosely, on the following scenario:

Mr. Leo Hooper, Jr. is thought to have contacted Mr. Harvey Peastone and arranged financing for the Boomerang. In return for his investment, Mr. Peastone would receive a necessary tax write-off for his business venture. Additionally, it is held by some

Leo Hooper Junior teaching a class, probably Hammer Calibration 103, most likely in the late thirties. Hooper initially wore a jacket and tie to instruct his classes. But he wasn't fooling anybody, and soon returned to his standard juvenile delinquent attire. ((Photograph courtesy of the Whitley Speale Collection of the Bone Lake Research Museum)

scholars (See: Bumpsteed, Jr.) that Mr. Peastone's intention all along was to use the Boomerang as a discreet alcoholic beverage delivery vehicle in his business enterprise.

FURTHER DEVELOPMENT OF THE FLUNGK Z-44 "BOOMERANG"

The initial design of the Boomerang, as it appears on the napkin in question, was for the time an extremely radical, even innovative, design. It must be remembered that in 1934, Ford tri-motors and Fokker airliners were the standard in air travel, although the Douglas DC-2, precursor to the venerable DC-3, would appear that same year. Although these were state of the art aircraft at the time, still they were rather staid airplanes. As cantilevered wing monoplanes, they represented a tremendous advance over the externally braced biplanes of just a few years before. Still, it must be stressed, these were conventional airplanes.

State of the art, 1934. Cockpit of a Ford C-4A, the military version of the Ford 5-AT Trimotor. (Photograph courtesy of the U.S. Air Force Museum)

would need men of tremendous vision and daring.

Little wonder, then, that when Mr. Peastone, canny New Englander that he was, was shown the drawing for the tailless Flungk Boomerang, he stated

Touted for the safety of its all metal construction, the Ford Trimotor became an early standard for passenger air service. It provided a fair representation of the state of the aeronautic arts when Professor Hermann Flungk had his startling revelation. This photograph shows a Trimotor in the service of the U.S. Marines in the late thirties. (Image courtesy of NASA Langley Research Center)

Alternative methods of flight were rarely even considered by the practical-minded aviation designers of the era. In 1934, a workable helicopter was still a figment of science fiction. Juan de la Cierva was just developing his autogyro in Spain, but this was virtually unknown in the United States. To reach its fruition, the concept of the boomerang aircraft

flatly that "he wasn't putting out one red cent unless he got a whole airplane with a real tail on it."

In this case, as so often happens, economic necessity had reared its ugly head, intruding upon the pristine beauty of the original design. Prof. Flungk swallowed his pride. Drafting pencil in hand, he extended the fuselage of the original Boomerang de-

119

The Douglas DC-2 first entered service in 1934, foreshadowing a revolution in air travel that would culminate in the legendary DC-3. It was in such a heady atmosphere of aeronautical innovation that Professor Flungk exercised his considerable aerodynamic acumen. (Photographs courtesy of the G. Eric and Edith Madison Collection of the Library of Congress.)

sign, and added a conventional set of stabilizers and control surfaces.

Another radical innovation of Prof. Flungk was a nose wheel on the original design. It is believed by reputable historians that this was originally installed to provide clearance for the rear facing propellor. It was now removed and replaced with a more conventional tail skid. In this instance, the tail skid was elevated to provide the additional prop clearance needed.

Through Prof. Flungk's extensive aviation contacts, an aircraft fuselage was finally procured. As it happened, years earlier, during the Blair Mountain mine wars, a Martin MB-1 biplane bomber had crashed near Logan, West Virginia. The hapless biplane had been deployed from Washington, D.C. in the only recorded instance of U.S. military aircraft being used against United States citizens. Sometime during the battle, it had careened through a farmer's chicken house, shedding its wings and engine. It finally came to rest in the farmer's hog pen, with its ruggedly constructed fuselage basically intact. Unfortunately, the wings had long since been used for hay tarps, and both engines had been sold to a local sawmill operator, but, the fuselage appeared basically intact.

Ever anxious to keep his research costs down, Prof. Flungk was able to negotiate the procurement of the fuselage. In exchange, he only had to replace the chicken house. And, by the most amazing coincidence, Professor Flungk discovered that the farmer, like Mr. Harvey Peastone, was also engaged in the alcoholic beverage industry. To show what a good sport he was, the Professor agreed to provide future limited delivery rights for the West Virginia farmer, at a professional rate.

One can only ponder the thoughts of the idealistic young men who had enrolled in Professor Flungk's Aviation Maintenance program, only to find themselves on a remote mountaintop in West Virginia, using their newfound technical skills to retrieve an airplane component full of chicken excrement, and then replace it by building a chicken house for a moonshining hillbilly.

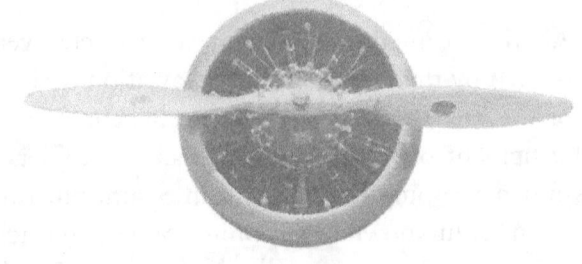

14. A COST-BENEFIT ANALYSIS OF AIRCRAFT PRODUCTION IN THE DEPRESSION ERA

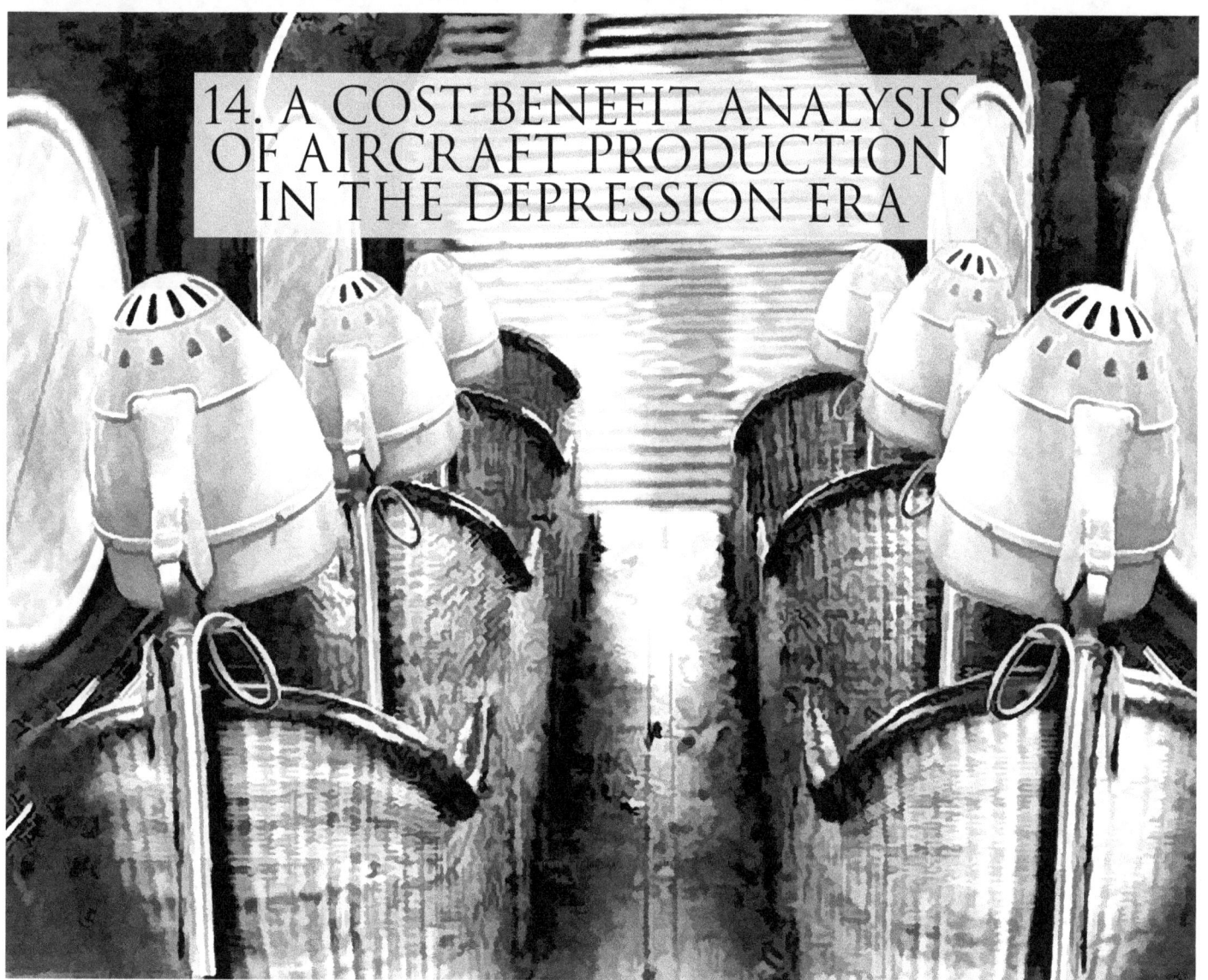

By the simple expedient of mounting six hair driers behind the passenger seats in the cabin of the Z-44 "Boomerang," Professor Flungk in one fell swoop provided additional training opportunities for the students of Z.S.ofA.M.&B.T., created a lucrative cash stream, and developed the world's first aerial beauty salon. (Photograph from the private collection of Dr. Eustis Bothomfieder)

Inspired by the success of his money saving venture into the hills of West Virginia, Prof. Flungk continued to institute further cost reducing measures. One of these was, of course, to utilize the Aviation Maintenance students as labor. This had the added advantage of providing them with real world hands-on experience. Another cost saving device was to utilize the discarded corrugated tin roofing from the chicken house for the wing and tail surfaces of the flying boomerang.

Prof. Flungk performed the necessary calculations (again on a napkin from Miss Kitty Kat's Top Hat Club), and determined that the additional weight from the heavy tin roofing was more than offset by the cost reduction over new aluminum aircraft sheeting. An additional cost-cutting feature was the utilization of the remaining red barn paint, which had been used to paint the chicken house, as the final color coat on the Boomerang fuselage.

A feature of the Flungk Z-44 Boomerang that has puzzled both the serious student of aviation history and casual observer alike, is the addition in some of the illustrations of what at first glance seems to be a large post-mounted helmet directly behind the pilot's station aft of the cockpit. A fact not known to many in the field of aviation history, is that this appendage is a Mark XIII Super Blow, a hair dryer that also was designed by Prof. Flungk. For years its purpose has remained hidden. Its usefulness is doubtful, unless one takes into account the economic factors involved in operating a technical school in the thirties. In addition to the pilot mounted hair dryer, there

A Martin MB-1 bomber. It was this type of aircraft from which Professor Flungk procured a fuselage to build his first model of the "Boomerang" series of aircraft, the Flungk Z-44 "Mark I." Inset: The Martin MB-1 which provided the fuselage for the Flungk Z-44 "Boomerang," shown shortly after its crash in September, 1921.

were an additional six hair dryers in the cabin section, making the Z-44 in effect, if not in actual fact, the world's first aerial beauty salon. The economic benefit becomes obvious when the tuition factor is considered. Now that the Z-44 served a function for the beauty school, Prof. Flungk could funnel funds from the beauty school tuition into the Boomerang slush fund. Plus, it would make a marvelous display for the beauty school to showcase on Parent's Day. Ever the entrepreneur, Prof. Flungk even developed a degree in aerial beauty application, thus creating the world's first beautician qualified flight stewardesses.

EVOLUTION: THE DREAM SLOWLY TAKES SHAPE

Once the chicken coop project in West Virginia was completed, the student mechanics returned to the school hangars. No rest was in store for them, however. They immediately went to work on the Boomerang project, which progressed at a frantic pace.

At least, it progressed at a frantic pace for Z.S.

of A.M. & B.T. students. It turned out that roofing the chicken house had provided valuable on-the-job training to the aviation Maintenance students, since the Boomerang's wings were made of the same material as the roof of the chicken house. Throughout the winter months of 1934, these hapless students groused and moaned, lamenting the fact that they had actually paid tuition money for the privilege of working around the clock in an unheated hangar on a project of such questionable significance. Of course, they were still in the midst of the depression, they had already paid in advance for their cafeteria meal tickets, and Prof. Flungk's famous "No work, no eat!" policy provided a strong work incentive.

Summer came and went. Then another summer came and went. The Boomerang slowly took the recognizable shape of a flying machine in the back corner of the hangar. Just as slowly, the "Water Whippet," a derelict that was sitting in the Zanesville hangar when Professor Flungk leased the facility, disappeared as the engines and various other parts were stolen from it, leaving little behind except an empty teakwood hull and some rotting linen.

Even with the repeal of Prohibition, the corn liquor business in New Hampshire remained strong, and Harvey Peastone kept the money rolling in. Construction was agonizingly slow, as is often the case with slave labor. Since the students weren't being paid, they had little incentive to exert themselves. Even this dovetailed nicely with the Professor's plans. The longer the construction took, the longer that Prof. Flungk could charge tuition fees,

and milk Peastone for construction money.

Although everyone concerned did all that they could to stave off the conclusion of the project, eventually the end loomed near. And, as the strange craft neared completion, an unsettling new question occurred to those associated with the Boomerang project.

If they actually finished building this monstrosity, would somebody have to fly it?

Emblematic of the heady atmosphere of discovery and innovation which surrounded Professor Flungk's development of the Flungk Z-44, was the invention of Juan de la Cierva, the Autogyro. His Cierva C-6, (pictured above, and inset, left), while never commercially successful, was widely imitated around the world. (Insets from left: Cierva C-6, Russian KASKR-1, Japanese Kayaba Ka-1, and U.S. Pitcarin autogyro.)

15. THE DREAM BECOMES A REALITY

"In the grand drama that is life, each of us takes center stage but for one brief moment, yet that moment shall be the one that defines us for all eternity. These few words may have constituted the thoughts that passed through the heated mind of young Leo Hooper Junior, as he strode across the dusty tarmac, mounted his gallant steed of the sky, and soared off into aviation history."

From: Whackos, Crackos, Sickos and Psychos--A Psychological Study of Fruitcakes and the Nutcases who Warped Them by Dr. Waldo von Heinkerblonker, M.S., PhD, L.S.M.F.T. Out of print.

"The choice of Leo Hooper, Junior as test pilot for the Z-44 had ample precedent in the annals of aviation history. The Montgolfier Brothers used a sheep, a rooster and a duck for their first balloon flight; Flungk used Hooper."

They Might Have Been Giants: Misadventures, Blunders, & Colossal Failures in Aviation by Halloway Bumpsteed, Jr.

Business side of the Z.S. of A.M.&B.T. hangar. In this photograph from the late thirties, the Flungk Z-44 "Boomerang" is in all likelihood inside, undergoing its final tweaks before its historic ascension. (Photograph courtesy of the Whitley Speale Collection of the Bone Lake Research Museum)

9. MARCH 13, 1938:
THE FATEFUL DAY DRAWS NEAR

At last, it seemed, there was nothing else to be done to put off the inevitable. Harvey Peastone, more than anxious to see some results from his investment, had angrily announced that he was getting on the next bus out of Chillblain and heading for Zanesville, Ohio. It was past time for payback. From his point of view, the sooner Flungk and his cronies got his contraption in the air, the sooner his money would come rolling back in. Prof. Flungk, for the most part philosophical, still hated

to see a good thing come to an end. The Z-44 had been a good cash cow, perhaps the best he had ever concocted. And yet, who knew? Maybe the thing would actually fly!

On the morning of March 13, 1938, as Harvey Peastone stepped down from the bus onto the cold concrete platform of the Zanesville bus terminal, he was met by Leo Hooper, Jr. Although Leo drove him around the countryside for hours, pretending to be lost, eventually he had no choice but to drive Mr. Peastone to the school. There Peastone was met by Prof. Flungk, who greeted him warmly. Anxious again harangued the Professor to take him to see the aircraft. However, Flungk was able to convince the financier to join him at Miss Kitty Kat's Top Hat Club.

Here, events become somewhat muddled. Indeed, there is nothing to indicate what had transpired until ten days later, when "Slappy's Flying Service," the aviation fuel supplier at Zanesville Municipal Airport, delivered 43.2 gallons of 80 octane aviation fuel, and one fifty-five gallon barrel of sixty weight aviation oil, to the Zanesville School of Aviation Maintenance & Beauty Technology. The bill for this service, below, was included with the 37 separate items that the author received from the

Mr. Harvey Peastone, the financial genius behind the "Flying Boomerang" project, whose economic theories left their mark on the aeronautical design process.

One of the items released as a result of the author's FOIA search was the infamous "Slappy's" fuel receipt, featuring Slappy's famous motto, "Slappy's not happy 'til you're slap-happy!" This receipt is also notable for being the single largest oil sale in Slappy's history, 220 quarts of Veedol 140 weight aviation oil. It isn't known whether or not this outstanding bill was ever paid, although its resolution in favor of Slappy's Flying Service seems highly doubtful.

to see his new aircraft, Peastone insisted on visiting the hangar. The Professor demurred, citing the late hour. Flungk suggested that instead of visiting the hangar, they tour the Beauty School. Here, Peastone was immediately distracted from his mission by the charms of the beauty students, as Flungk had planned. When classes finally ended, Peastone once

USAF Materiel Command.

No aircraft is specified, but, as Z.S. of A.M. & B.T. possessed only the now derelict and engineless "Water Whippet" and the newly constructed "Boomerang," it must be surmised that the fuel and oil was indeed intended for the Z-44.

On the date of the Slappy's document, March 23, 1938, records of the National Weather Service, show that a cold front had passed across the Zanesville Municipal Airport in the early hours, pushing out a late winter squall line. This would in all likelihood have resulted in a marvelous sunrise, as the giant glowing orb of the sun rose magnificently in the eastern sky, breaking through the low lying clouds of the recent storm, and sending brilliant shafts of early sunlight shooting across the heavens. As the radiant sunlight struck the minute water droplets still dispersed throughout the lower atmosphere, there may have even been a rainbow. Thus would nature itself create a dramatic backdrop for this glorious culmination of years of planning, building, heartache, sweat, and tears: The first flight of the Flungk Z-44 Mark I "Boomerang," the revolutionary aircraft that would change forever the very heart and soul of aviation.

Unfortunately, no one was present at the airport to witness if this was indeed the case. Due to the almost continual festivities of the last ten days since

The Flungk Z-44 "Mark I" Boomerang in front of the Z.S.ofA.M.&B.T. hangars, probably in the early fall of 1937. This would, of course, have been before the historic first flight. Inset: Z-44 wih optional Flungk "SuperBlow" hair drier installed. (Photographs courtesy of the Whitley Speale Collection of the Bone Lake Research Museum)

the arrival of Harvey Peastone, no one at the Zanesville School of Aviation Maintenance & Beauty Technology was out of bed before noon. When the Z-44 was finally rolled out of the hangar somewhere around two PM, it was at the insistence of Mr. Peastone, who had threatened Flungk with a visit from some of Peastone's Italian business associates from New Jersey, if he didn't produce the project airplane, and quickly. The Aviation Maintenance students were rousted from their sleep, and they groggily stumbled out to the hangar, pushed back the hangar doors, and rolled the Flungk Z-44 Mark I "Boomerang" out onto the tarmac for its historic first flight.

By this time, a contrary nor'easter had pushed the previous night's storm back over the airfield and stalled, creating a vicious mix of rain and sleet with wind gusts of up to forty knots. The Z-44 was in a very real and imminent danger of being blown away in the wind. Therefore, the decision was made to call out the entire student body of the Z.S. of A.M. & B.T., which included the Beauty School Students. The School receptionist was sent over to the Blue Moon Motor Inn to wake Hilda Grackle. After fifteen minutes of pounding, the receptionist got Hilda out of bed. Together they roused the Beauty School students. Due to the intense social activities of the night before, nearly all of the Beauty Students, who moonlighted at Miss Kitty Kat's Top Hat Club, were in a state of deep repose at the Club, still attired in their "Kitty Kat" costumes. Still, when called upon they responded, if somewhat groggily. This would explain why, if an unsuspecting traveller was driving down the airport road on that afternoon of March 23, 1938, he would have seen, through an almost impenetrable haze of greasy smoke, an incongruous aircraft being held down by a dozen young women scantily clad in Kitty Kat costumes, shivering in the blast of rain and sleet, as they prevented the Z-44 from being blown away in the wind.

While the Kitty Kat girls bravely held on to the bucking airplane, a specially selected cadre of the best and brightest from the Aviation Maintenance side of the School, primed, cleared, pumped, and propped the cantankerous Whutley-Spitsworths in a glorious but seemingly futile attempt to get both engines running at the same time. As usual, no sooner would one Whutley-Spitsworth cough, belch, and blat to life, than the other would promptly quit. Each time this happened, both engines would send huge clouds of greasy grey smoke billowing up into the air, which would thankfully be whisked away by the gusting winds.

In addition to her duties as the Cigarette Girl at Miss KittyKat's Top Hat Club, Amanda Smilch (née Amanda Gundersen) also worked on the maintenance side of Z.S.of A.M.&B.T. She is shown here cutting parts for the Flungk Z-44 "Boomerang," probably in 1937. (Photo courtesy of Dr. E.C. Whimpington)

THE GRUNDT-SMILCH TAPES

The events of March 23, 1938 as they related to the alleged flight of the Flungk Z-44 "Boomerang" remain one of aviation history's great mysteries. Today, almost no records exist to illuminate the events of that somewhat fateful day, with one glaringly marvelous exception. Fortunately for posterity, in 1996 Willard Grundt, then a student at the Mason County School of Anthropology Technology and Phlebotomy at Point Pleasant, West Virginia, was assigned to write a term paper on the history of Vocational-Technical institutions within the region. In the process he came across a Mrs. Amanda Smilch, then residing at the Sunnydale Convalescent Home in Marietta, Ohio. Although incoherent at first, Grundt says that Mrs. Smilch became more responsive once her restraints were removed. Grundt visited Mrs. Smilch on several occasions during the late winter and early spring of 1996. In her brief moments of seeming lucidity, he was able to interview her about her days at the Zanesville School of Aviation Maintenance & Beauty Technology. Luckily, these events were recorded on audio tape, which Mr. Grundt has graciously allowed access to. What follows is a portion of those transcribed recordings.

Grundt: Hello, Amanda. I'm Willard Grundt. Do you remember me? I was here to visit you last week.

Smilch: Riley? Is that you, Riley?

(Mrs. Smilch is evidently referring to her late husband, Riley Smilch.)

Grundt: No, Amanda, it's me, Willard. I'd like to talk to you again.

Smilch: Make them let me up, Riley. I promise I won't bite anybody.

(The conversation goes on in this vein for some time. As Mrs. Smilch drifts in and out of coherence, Grundt questions her about her time at Z.S. of A.M. & B.T.)

Grundt: What was your function at the Technical school?

Smilch: Thanks for getting those cuffs off of me, Riley. Could you get me a smoke?

Grundt: Were you a student under the late Professor Flungk?

Smilch: Hah! The only time we ever saw the Professor was at the Club.

(**Mrs.** Smilch refers to Miss Kitty Kat's Top Hat Club, located near the Technical School and scene of many school ceremonial functions. EDITOR.)

Grundt: Were you at the club often?

Smilch: Riley, whatever has happened to your mind? You were there, too! It's where you proposed to me!

Grundt: I'm starting to remember now. Why don't you refresh my memory, though. It's still a little fuzzy.

Smilch: Well, silly, you must remember that all of us girls in the Beauty School worked nights at the Club. It was the only way we could make ends meet. I was a cigarette girl, remember? I liked that because my costume wasn't as skimpy as some of the other girls.

Amanda Smilch, at left, takes a break from her duties as cigarette girl at Miss Kitty Kat's Top Hat Club. The two girls seated with her are unidentified, although it has been postulated that the girl on the right is Hilda Grackle's daughter, Esmerelda. (Image courtesy of the private collection of Dr. Manfredd von Goetzzenberger of Das FleugelWerkes.)

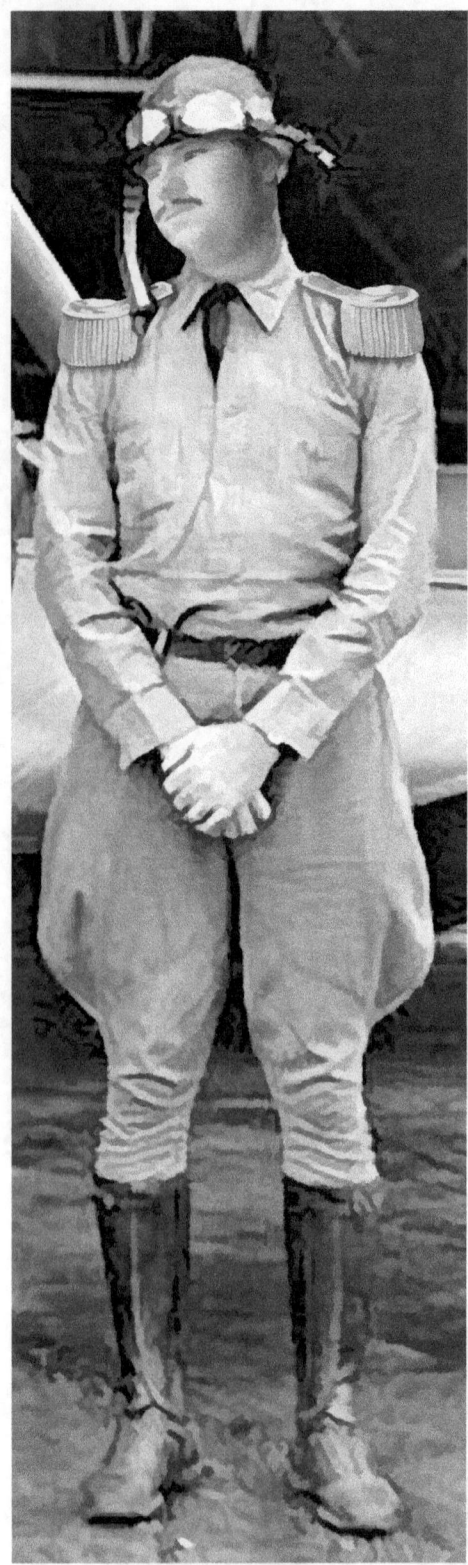

Although this snappy uniform wasn't available for the inaugural flight of the Z-44, Leo persuaded Harvey Peastone to spring for it in anticipation of the new "Test Pilot School." Leo was so excited with his new uniform that he commissioned the above oil painting in late 1939. (Artifact courtesy of Dr. E.C. Whimpington)

Grundt: What was Professor Flungk like? Nobody seems to know much about him.

Smilch: (Laughs) I'm surprised you would even care! After all, he kicked you out of school right after you paid your tuition! I know I used to want to just kill him! Sitting there at his little front row table, snapping his fingers and saying "Cigaretta! Cigaretta!" and when I walked over, he would snap my garter and start cackling! I used to have a permanent welt the size of a silver dollar on my thigh! He gave me the willies, what with that dent in his head and all. And that juvenile delinquent he kept around, what was his name?

Grundt: Would that be Mr. Leo Hooper, Junior, the Assistant Instructor?

Smilch: That's it! Leo! That skinny little suck-up! He was Leo alright, only in the Club he was always saying, "No, no, don't call me Leo. I am Raoul! Call me Raoul!" Sitting there, lighting the Professor's 'cigarettas' and ordering his drinks for him. All dressed up in his fake uniform, gold braids on his shoulders like a little tin-pot dictator! He might have fooled some people, but he sure didn't fool me! I remember the night they were going to launch that whatchamacallit thing of the Professor's- What was it, Riley? You remember. You were there!

Grundt: Are you referring to the Professor Flungk's Z-44 Mark I "Boomerang."

Smilch: That's it! That "Boomerang" thing! Hah! What a joke that was! The night before they were going to fly it, Leo and that Mafia guy from up north, what was his name, Gallstone?

Grundt: Do you mean Mr. Harvey Peastone, the New Hampshire distiller and shipping magnate who was a protege of Professor Flungk's?

Smilch: Peastone! That was his name! That fat little sawed-off twerp really thought he was God's gift to women! Had a hairpiece that looked like a cocker spaniel sleeping on his head! Anyway, they were all sitting around the Professor's table, whooping it up. And Leo was bragging all about how he was going to fly that "Boomerang" thing the next day. He was going on about how he could wear his welding goggles and shop coveralls just for tomorrow's flight, but now that he was a test pilot, they were going to have to buy him a real flying helmet, goggles and a flight jacket. Then Gallstone-

Grundt: Peastone.

Smilch: Gallstone, Peastone, whatever! You're always interrupting me, Riley! Anyway, Peastone started sputtering about how that wasn't in the budget, and they'd spent too much money already. And then- Oh, this is too much! (Laughter) Then Leo told Peastone not to worry because they were going to come out money ahead in the long run. (Laughter) This is the best part!

They were going to make a lot of money because Leo was going to teach a Test Pilot Class at the School! Can you imagine?

Grundt: Just barely.

WHERE IN THE WORLD IS LEO HOOPER JUNIOR?

Grundt: Then what happened?

Smilch: I'm not sure. I suppose Gallstone was getting fed up with Leo. It didn't take much of him to do it. Finally, Gallstone smacked Leo across the mouth. Then Leo jumped up and threw his martini in Gallstone's face. And all the while the Professor just sat there patting his hands in front of himself like he was trying to quiet them down, and mumbling something over and over in German. I think I passed out sometime after that, because the next thing that I remember was Hilda shaking us all awake and yelling that we had to go save the "Boomerang," whatever that meant. So we all hustled over to the hangar, all of us freezing our little buns off because we were still in our costumes. The Aviation School boys were holding on to that whacky airplane, but they were about to lose it, you could tell. They had us grab onto it and hold it down so that the Aviation School boys could try to get the engines started.

Grundt: Where was the Professor during all this activity?

Smilch: Oh, he was there alright, strutting around like a banty rooster! Walking around in circles, spouting orders in German, and nobody there knew what on earth he was talking about! And him in a full length raccoon coat, while we girls all had so many goose bumps we looked like plucked turkeys!

Grundt: What about Leo?

Smilch: Leo? Hah! Mr. "Test Pilot!" Nobody could find him, you see. Everybody was saying "Where's Leo? Where's Leo?" We all knew that any time now the boys might get those engines started. At least we hoped that they would! Then somebody would have to fly that thing, which meant that we could all go back to bed! We all thought the pilot was supposed to be Leo. After all, the night before Leo, or should I say, "Raoul" sure was hot to trot, spouting off all that "test pilot" nonsense! Well, let me tell you, Gallstone was getting hot! No, he wasn't hot, he was steaming! I just know that he was beginning to think that the whole "Boomerang" project was a scheme cooked up by the Professor and Leo to take his money. If you ask me, it probably was.

Grundt: So where was Leo?

Smilch: I'm getting to that. (Laughter) This is just too good! Buchard Woolsey, one of the aviation

Actual photographs of the Flungk Z-44 "Boomerang" are extremely rare. However, the above artist's reconstruction is based upon materials known to be used in the construction of the aircraft, and is believed by most experts to be an accurate representation of the Z-44 on the occasion of the inaugural flight, with one small exception. It is not known whether the Flungk "Superblow" hair drier shown mounted at the pilot's station was actually installed on the day of the flight. (Artist's reconstruction courtesy of Halloway Bumpsteed, Jr.)

students, went into the break room to get a Coke out of the machine. Only the machine took his nickel and didn't give him a Coke. Now, Buchard was just a big dumb farm boy, but throwing hay bails all that summer had made him fairly stout. He smacked the machine so hard it rocked back against the wall. Nearly knocked it over. And the machine squealed at him, "Ow! Watch it, you're hurtin' me!" Buchard's hair stood up on end, and he just about jumped out of his skin! He ran back outside screaming, "The Coke machine is haunted!" "What do you mean?" somebody asked him. "It was talkin' to me!" he said, and everybody cracked up. Everybody that is, except Gallstone.

Grundt: Do you mean Peastone?

Smilch: Gallstone, Peastone, what's the difference? As far as I'm concerned, a rock is a rock! You want to hear this story or not?

Grundt: Don't mind me. Please, go ahead.

Smilch: So Gallstone got this look, you could almost see a light bulb go on over his head, just like in the cartoons. He got a couple of the boys and went into the break room. A minute later you could hear the most awful caterwauling in there, you'd have thought they were killing somebody! Then out came Gallstone. The two boys were right behind him, dragging Leo between them!

Grundt: So they found Leo?

Smilch: He was hiding behind the Coke machine the whole time! Leo was screaming, "No! No! Don't put me in that thing! I don't want to go! You gotta send somebody else! Anybody!" I guess

he taught himself that in his Test Pilot School. Of course, he still had on his fake soldier get-up from the night before. You would have thought at least that he would have some respect for the uniform. It was a disgusting display, let me tell you. Another hour went by and they still hadn't gotten both engines running at the same time. They would get one running, then start the other one, and the first engine would die. Then they would start the process over again. All this time, those engines were belching out this nasty old greasy black smoke. Half the girls were puking their guts out, while the other half were holding on to that "Boomerang" for dear life. The wind was still howling, and by now our goosebumps were as big as golf balls.

Grundt: What was Leo doing all this time?

Smilch: He tried to run back inside and hide behind the Coke machine again, but the boys just went in and dragged him back out. After a while he just sat against the side of the hangar, whimpering. Feeling sorry for himself, I suppose. Finally, he got this crafty look on his face. I think he figured that they couldn't get the darn thing to run, and he wasn't going to have to fly it after all! Pretty soon he stood up, and it wasn't very long before he was running around shouting orders, even though he looked so stupid with his fake penciled-on mustache running down his chin in the rain. Leo was getting cockier by the minute, and I could tell he was just getting ready to say, "No, no! Call me Raoul!" Then, wouldn't you know it, the Tech School boys finally got both engines running at the same time.

Purportedly one of the propellors from the ill-fated Flungk Z-44 Mk I "Boomerang." This object is in the private collection of Dr. Manfredd von Goetzzenberger of Das FleugelWerkes; however, its provenance is highly doubtful. (Photograph courtesy of Ad Meskins.)

17. THE LIMITS OF VALOR

Although this photograph is heavily damaged, it retains its historical significance as one of the handful of photographs featuring Professor Hermann Flungk (center, with boomerang) in a rare public appearance. The occasion is unknown, as is the object in the background, which may be an aircraft, though it is doubtful that it is the Z-44 Mark I. Also identifiable are Hilda Grackle, (aka "Miss Kitty Kat"), at the far left, Mr. Harvey Peastone, third from left, and a young Leonidas Hooper, second from right. (Courtesy of the Whitley Speale Collection of the Bone Lake Research Museum)

CREW ASSIGNMENT ON THE Z-44 "BOOMERANG"
(The Grundt-Smilch Tapes, Continued)

Grundt: What was Leo's reaction to all this?

Smilch: You could see the blood drain out of Leo's face. I had always heard of people's faces turning white, but I hadn't seen it until that day. He started to run, but Gallstone yelled, "Don't let him get away, boys!" Buchard Woolsey jumped down off the wing of the "Boomerang." I had never seen a big boy like that move so fast before. He tackled Leo, picked him up under one arm, and carried him up the stairs to the pilot compartment up in the nose.

Leo was squealing bloody murder. Buchard lifted Leo up over his head and stuffed him into the pilot seat. Leo was flailing his arms around, slapping Buchard all around his head, until Buchard balled up his fist and punched Leo point-blank right in the forehead. He knocked Leo out cold! Leo's head rolled back and his arms dropped down over the side of the airplane.

Buchard strapped Leo in and pushed the throttles forward. He jumped down off the stairs and pulled them back out of the way, yelling "Let her rip, ladies!"

I'm not sure why he was so wound up. I think he must have blamed Leo for his lost Coke. All of us girls still holding on to the "Boomerang" let loose at once and ran away from it as fast as we could in our cute little Kitty Kat pumps. The engines coughed and sputtered, and we all thought for sure that they were going to die. Then both engines backfired, belching out more huge clouds of black sooty smoke. The wings started to rock, and just when it looked like the wind was going to blow the whole mess away, the engines caught. The whole airplane started

Buchard Woolsey, a tough farmboy, wasn't above throwing his weight around, which was considerable. In the end, though, even Buchard Woolsey was no match for a Coca Cola machine. (Photograph from the private collection of Dr. Eustis Bothomfieder)

to spin around with this big "Whoosh! Whoosh!" sound that you could hear even over the screaming engines and the howling wind.

Just then the wind picked up and the "Boomerang" skittered sideways. It was rotating at a pretty good clip now, and we could see Leo's arms

fly out, just like he was on a carnival ride. The tail skid was scraping across the tarmac now, throwing up a huge shower of sparks. The "Boomerang" was headed straight for the hangar, and it was obvious Leo wasn't going to do anything about it. Some of the girls tried to grab it again, but they changed their minds real quick. They might as well try to grab a buzz saw in a tornado! So we all just stepped back and watched the show. By the time it reached the hangar it was whirling so fast it was just a blur. When it smacked into the hangar I knew even the boys over at Slappy's Flying Service on the other side of the field must have thought a bomb had gone off. There was a huge explosion of dust, smoke, flying glass and fabric off the airplane, but we could hear that the engines were still running. The winds blew the dust away pretty quick. When the smoke cleared, we couldn't believe our eyes! The tail of that contraption had come around at just the right instant for it to slam into the open hangar door. The tail was cut clean off, bringing the airplane to a complete stop.

The hangar door, with all its glass blown out, teetered for what seemed to us like forever, but it probably wasn't any time at all. Then a wind gust picked up the door like it was a sheet of cardboard and blew it across the airport. And wouldn't you know it, the same gust of wind blew what was left of the "Boomerang" out away from the hangar just like it was planned that way! Well, those engines were still howling like they would blow apart at any second. It wasn't any time at all before the "Boomerang" started spinning again, only now that the tail had been cut off, it really sped up. First there was a low thumping sound, but it kept getting louder as the airplane spun up, and climbing higher up the scale until we all had to hold our ears. I've never heard anything like it, before or since. Now the wind was blowing the cursed thing across the tarmac at thirty or forty miles an hour. All of a sudden it jumped up off the ground and started to climb up into the sky! About that time Leo must have come to, because even over all that noise, we heard the most pitiful scream come out of that thing. It sounded to us just like Leo was being carried off by demons. It probably felt just like that to Leo, too. We watched it for a few seconds until it disappeared into the clouds. Some of the girls started to cry. One of the boys remarked that he was sure going to miss old Leo. Buchard said that if Leo didn't make it back, well, it couldn't have happened to a nicer guy.

A SIGN FROM ABOVE:

Grundt: But surely Leo didn't simply disappear?

Smilch: No such luck. Just then there was a tremendous "Boom!" followed by a loud clanking

but it wasn't. We could still hear the whooshing of the wings and Leo's pitiful screaming. Somebody yelled "Look! Over there!" and pointed across the airfield. On the other side of the runway, we could see what was left of the "Boomerang" break out of the clouds and spin down just like a giant whirligig seed, trailing a corkscrew of smoke and sparks. It hit the ground with a loud "Whoomph!" and bounced about five feet back in the air, then came back down

Amanda Smilch was justly proud of her Flight Beautician Certificate, which was issued to her in May, 1937, before her marriage to Riley Smilch. Of passing interest is the signature of the Chairman of the Borad of Regents, "Mme. Hilda Grackle," otherwise known as "Miss Kitty Kat." (Artifact courtesy of Halloway Bumpsteed, Jr.)

noise from out of the clouds. We could still hear engine noise, only not as loud now. A few seconds later, there was another loud "Boom!" and some more clanking for a few seconds. Then the clanking stopped. I guess both engines finally blew up. After all that noise, it seemed like dead silence,

and rolled into the snow fence. For a second it got deathly quiet. Even the wind died down. Everybody stopped talking all at once. It was creepy. All you could hear was Leo, howling like a stuck pig.

Grundt: So you mean to say after all that, Leo was still alive?

Smilch: No, you moron, it was only his ghost squalling like a boar hog in heat! Of course he was still alive! Do you think a dead man could make that much noise? Well, the quiet only lasted for a second, then everybody was yelling and screaming all at once. From over at Slappy's Flying Service, we could hear a siren. Some of the students started running across the tarmac toward the wreck. Four or five of the girls took off to rescue Leo, waving their arms in the air and yelling "Oh! Oh! Poor Leo!" and "Don't worry Raoul, we're coming! We'll save you!" I couldn't believe it. Now they were calling him "Raoul!" I almost lost my groceries right there. Of course, as soon as they left the tarmac, their Kitty Kat pumps got stuck in the mud and they all landed flat on their tushes. A couple of the more sensible boys ran into the hangar to start up the school's

old Model-T truck. And then the darndest thing happened. I was watching all this action, everyone running around like a bunch of chickens with their heads cut off. Over the runway I could see what looked like a couple of pigeons drop down out of the clouds, fluttering around each other like they were dancing in the air. Buchard was running toward the wreck, and I saw him look up and see them about the same time I did. They landed smack dab in the middle of the runway about thirty yards in front of Buchard. Well, Buchard ran over and picked them up. He looked at them for a minute, then he threw back his head and started laughing. Even with all the noise, I could hear him cackling like a maniac. Guess what those two dead pigeons were?

Grundt: Beats me. The only thing I can think of, were they two dead pigeons?

These epaulets, now residing in the National Museum of the People's Democratic Republic of Schmuicklund, were purported to have belonged to an unknown Allied airman in the Battle for Schmuicklund. It is highly likely that these are the epaulets worn by Leonidas "Raoul" Hooper, Jr. How they came into the possession of the Nation Schmuick Museum is the subject of a later chapter. (Photograph courtesy of National Museum of the People's Democratic Republic of Schmuicklund)

Amanda Gundersen proudly models her new Flight Beautician's uniform on graduation day, May 14, 1937.

Smilch: They were Leo's gold braid epaulets! They must have been thrown off of him in the clouds, and then just fluttered down onto the runway, as pretty as you please.

RETURN OF THE HERO:

Grundt: What was Leo's condition after all of this?

Smilch: By this time, those mechanic students had gotten the Model T truck started. When they finally got across the runway to what's left of the "Boomerang," Leo had gotten out. He was trying to walk, but all he could do was spin around in circles. Finally a couple of the boys got hold of him. They immediately let him go, because he was covered in his own vomit. Finally one of them got a blanket out of the emergency kit in the truck and they tackled him and wrapped him in the blanket. Then they threw him in the back of the truck. He was still trying to spin around, and it took four of them to sit on him and hold him still.

Grundt: How long was he in the hospital?

Smilch: Hah! Who was going to spend any money to send him to the hospital? As it turned out, there wasn't really anything wrong with him anyway. His head was just in a spin after that wild ride of his. Now you would think that after all that, he would take a few days off, or at least calm down some. Oh, no! Would you believe it? That very night he was back at Miss Kitty Kat's Top Hat Club, strutting around and bragging about his big test flight!

Grundt: I would think that Peastone wouldn't be very happy about the loss of his investment.

Smilch: Are you kidding? It took three guys to hold him off when they brought Leo back to the hangar! If he'd have gotten loose, I swear that he would have torn Leo's head off! The Professor and Leo finally got him calmed down later in the day. They convinced him that the wreck of the "Boomerang" was just a minor setback. Leo told him it would all be rebuilt in a couple of weeks. He told Gallstone to take the night bus home, and they would call him when it was ready to fly again. "Actually," says Leo. "The test flight today proves that the Professor's idea was right on the money!" "You're telling me that this was a big success?" says Gallstone. "You crashed our airplane!" "Well, that's nothin'!" says Leo. "Us test pilots crash airplanes all the time! It's how we make our livin'!" Leo and Professor Flungk talked Gallstone out of another $5,000 to rebuild the "Boomerang." "That's just a drop in the bucket!" says Leo. "You're gonna make millions!" That evening they put him on the bus back to Chillblain. The very next day Leo phoned in an anonymous tip to the FBI about Gallstone's "business" back in New Hampshire. The FBI and the ATF kept Gallstone so busy he didn't have time to worry about his little airplane project.

Grundt: How did the Professor take all of this?

Smilch: Who ever knew about the Professor? The night after the crash he was back at his little table in the Club yelling "Cigaretta! Cigaretta!" and cackling every time he snapped my garter. I would've slapped him up the side of his dented head with my cigarette tray if it wouldn't have cost me my job!

18. DID THE Z-44 ACTUALLY FLY?

AN HISTORICAL PUZZLE
WITH OMINOUS IMPLICATIONS:

The Flungk "Boomerang" is never to be confused with the Australian Commonwealth Aircraft Corporation CAC "Boomerang," pictured above, which held the production contract numbers CA-12 through CA-19 during the war years of 1942 through 1945, and which actually flew.

"The events of March 23, 1938 will go down in history as an achievement of almost immense proportions. Never before had man attempted to emulate the graceful flight of the aboriginal boomerang with such audacity. Unfortunate though it is that the Z-44 Mark I Boomerang was destroyed on its maiden flight, the aeronautical principles and theories of Professor Flungk were demonstrated in a manner never before witnessed in the brief history of aviation. The heroic airmanship of Leonidas Hooper, Junior, on that momentous day, has placed Professor Herrmann Flungk at long last in his rightful niche in the vaunted pantheon of aircraft designers."

> From: *They Dared to Fail:*
> *Notable Detours on the Airways of History*
> by Dr. Eugene C. Whimpington

"Yes, yes, in a sense, you could say that it flew. However, with a forty knot wind, you could fly a sheet of plywood!"

> From: *They Might Have Been Giants:*
> *Misadventures, Blunders, & Colossal Failures in Aviation*
> by Halloway Bumpsteed, Jr.

Even to this day, the contentious dialogue over whether the Flungk Z-44 "Boomerang" accomplished a successful test flight before its untimely destruction rages unabated. Serious scholars of aviation history on both sides of the debate have vociferously defended their viewpoint, at times coming to blows, though only rarely requiring police intervention. There remains surprisingly little contention as to whether the "Boomerang" actually left the ground. Few scholars seriously dispute the

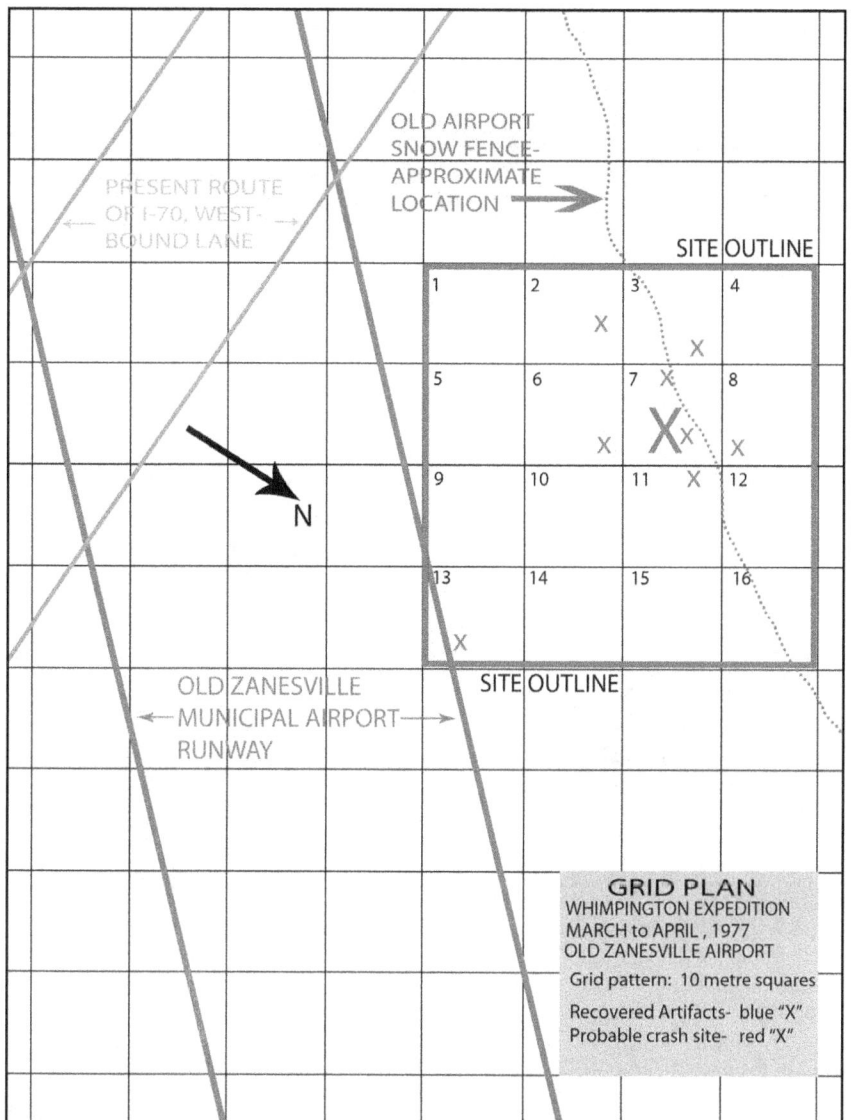

Grid plan for the 1977 Wimpington Expedition. The expedition team was able to distinguish the approximate location of the crash site by the recovered artifacts, each designated by a small "x" in the above diagram. (Graph courtesy of Dr. E.C. Whimpington)

least to the gullible, to corroborate the testimony of Mrs. Smilch. These included wood fragments that were consistent with aeronautical building practices in use at the time of the manufacturing of the Z-44, including a badly decomposed section of spar, and a fragment apparently used in rib construction. Whimpington also found a badly damaged fragment of corrugated metal that matched chicken coop roofing produced in the years immediately preceding the construction of the Z-44 Mark I; a specimen of fibrous material, which was later confirmed by laboratory analysis to be horse hair, most likely used in the seat stuffing of the Z-44; a corroded cast metal artifact, most probably a gear mounting bracket, which was still coated with paint residue containing chemicals consistent with the ingredients of red barn paint of the era; and, last, but perhaps most tellingly, a corroded transformer which matched a transformer in the housing of the only remaining Flungk Mark XIII "SuperBlow" hair dryer, which is currently on display at the Museum of Beauty Science in Waukegan, Illinois. Additionally, using the latest state-of-the-art laser scanning techniques available at the time, Whimpington made a microscopic inspection of the area surrounding the old Zanesville runway. He claims to have located minute gold fibers that purportedly came from the epaulets of Leo Hooper, Junior. However, even staunch advocates of Professor Flungk find this last claim a trifle hard to defend. Perhaps in the future, new technologies will allow this minor dispute to be scientifically settled. Neither is there any genuine argument over whether the Z-44 was destroyed in the ensuing crash. After the ill-fated test flight, the hapless boomerang was towed back into the hangar, where Flungk continued

eyewitness account of Amanda Smilch, even though her mental condition at the time of her interviews was somewhat questionable. Additional corroboration was provided by Dr. Eugene C. Whimpington in his 1977 archeological expedition to the crash site, which was filmed by Horrific Productions, and subsequently aired on various community access cable channels under the title *Finding Flungk's Folly: In Search of the Z-44 Boomerang*. Whimpington's team recovered remnants of the snow fence, and they were able to determine its approximate position at the time of the crash. Following the line of the old snow fence, they discovered various items that seemed, at

ITEMS RECOVERED BY THE 1977 WHIMPINGTON EXPEDITION

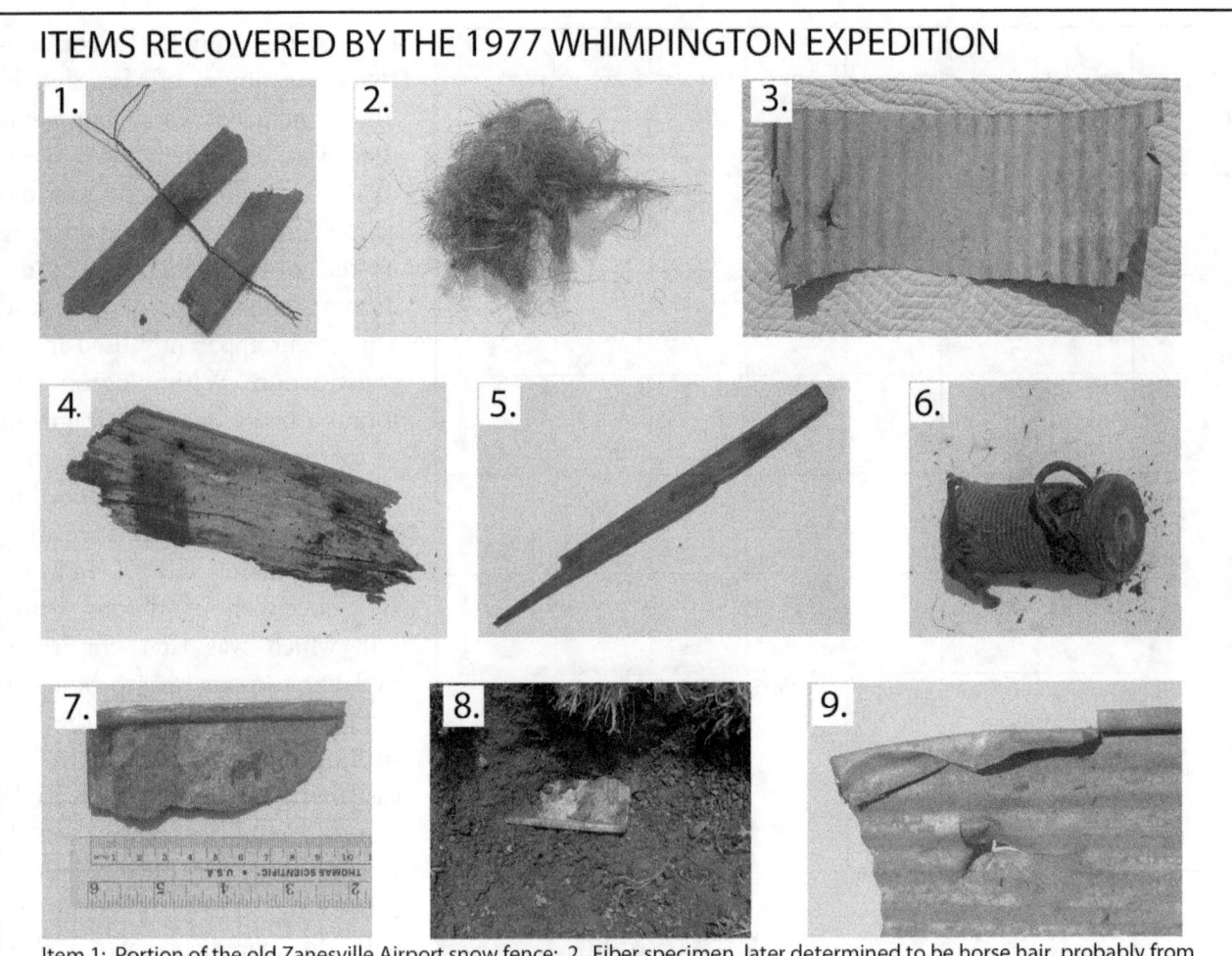

Item 1: Portion of the old Zanesville Airport snow fence; 2. Fiber specimen, later determined to be horse hair, probably from the pilot's seat of the Z-44; 3. Corrugated metal wing covering section; 4. Wing spar remnant; 5. Rib section; 6. Flungk Mark XIII "SuperBlow" transformer; 7. Cast metal fragment; 8. Cast metal fragment in situ, before removal for cataloguing. Notice the paint remnants, which were later determined to be consistent with barn paints of the period. 9. Detail of the corrugated metal wing covering section, showing evidence of the extreme trauma to which it had been subjected. This was considered by Whimpington to be substantiating documentation that it was indeed a portion of the wing covering from the Z-44.

The above illustration originally appeared in the definitive study of all things Flungkian, *They Dared to Fail: Notable Detours on the Airways of History,* by Dr. Eugene C. Wimpington. It is reprinted here by the gracious permission of Dr. Whimpington.

to utilize the cabin as a simulator for his aerial beautician students. For years Flungk continued to churn out flight-qualified beauticians, but unfortunately, a viable commercial market for these services never fully developed. However, it has been reported that a 1947 Flight Beautician certificate, issued to a "Miss Clara Dicionnara," and supposedly signed by the Professor, was recently offered on Ebay. At the time of this writing it had reportedly been bid up to $2.36. It is also generally accepted that the inclusion of the person of Leonidas Hooper, Junior in the aircraft qualified the flight as "manned." Detractors of Professor Flungk's role in aviation history are quick to point out that Mr. Hooper had to be physically inserted into the aircraft by one of the Technical School students, a Mr. Buchard Woolsey. Yet, even these naysayers must admit that, in many other aerospace operations, such as space flight, for example, the crew in their bulky suits must often be assisted by others in both entry into and egress from the respective vehicle. Quite literally the effort of thousands of individuals are often required to insure the successful outcome of a single mission. It is quite small-minded indeed to disregard the efforts of so many like Mr. Woolsey, so crucial to the successful outcome of a mission, yet who must stand in the shadows on the sidelines when the accolades of fame rain down on those in the spotlight. For every

THE PRICKLY ISSUE OF "CONTROLLABILITY"

History has not been unkind to Mr. Leonidas Hooper, Jr., although there are those who claim that his greatest attribute was being in the wrong place at the right time. (Photograph from the private collection of Dr. Eustis Bothomfieder)

Inevitably, in debates such as these, the issue of "controllability" raises it's ugly head. Was the "Boomerang" capable of controlled sustained flight, or merely responding to the whims of its environment, much like a leaf on the wind? This question, more than any other, has haunted scholars for decades. This is precisely because the concept of conscious control is generally accepted as an essential characteristic of an aircraft as it is commonly defined. That there were numerous attempts at powered flight before the Wright Brothers, some of which achieved a very respectable height and distance for the time, is well known. The Wright Brothers, however, were the first to operate a machine capable of being turned, therefore making it capable of powered, sustained, and controlled flight. Thus they are rightly regarded as the first to fly, a regard based heavily upon the controllability of their aircraft. Others had achieved powered flight before them; others had achieved sustained flight; theirs is rightly regarded as the first controlled flight. Regarding the events of March 23, 1938, one thing is clear: The definition of

Leonidas Hooper, Junior, riding in a convertible in a ticker-tape parade, there is at least one Buchard Woolsey following behind with a broom and push basket. This state of affairs is even more reprehensible for the fact that Mr. Woolsey is unable to defend himself. His life ended tragically only a few short months after the flight of the Flungk Z-44 Mark I "Boomerang," when a soda machine in the break room of the Zanesville School of Aviation Maintenance & Beauty Technology tipped over, crushing him. A dark spot in the concrete floor of the break room still marked the scene of this unfortunate incident as late as 1956.

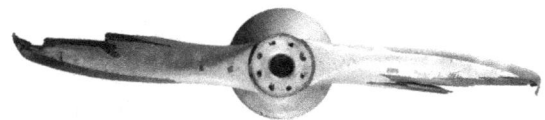

Unscrupulous historians still inveigh loudly against Buchard Woolsey, even though he hasn't been around to defend himself for some time. (Photo courtesy of Dr. E.C. Whimpington)

Wright Brothers Type "A" Flyer at Fort Meyer, Virginia, 1909. The Wright Brothers' claim to fame was based on their ability to control their aircraft. (Photograph courtesy of the Harris & Ewing Collection of the Library of Congress.)

flight, like beauty, seems very much to be in the eye of the beholder. Dr. Eugene C. Whimpington, in his seminal study, *They Dared to Fail: Notable Detours on the Airways of History*, sidesteps with characteristic obtuseness: "The concept of controllability does not exist in a vacuum; in point of fact, it is well known that controlled flight is impossible in a vacuum; this is incontrovertible. I prefer to approach the question of controllability in terms of 'outcome-based' parameters. This rational, yet immensely practical exercise, quite simply stipulates that controllability results if the anticipated intent corresponds to the subsequent development. In layman's terms, if you shoot an arrow and it hits the bullseye, then you win the archery contest. No one, least of all Bumpsteed, Jr., can argue that the intent of Leonidas Hooper et al on March 23, 1938 was to aeronavigate the Flungk Z-44 Mark I "Boomerang" from point A to point B, a maneuver which undoubtedly occurred."

Bumpsteed, Jr., in his monumental tome, *They Might Have Been Giants: Misadventures, Blunders, & Colossal Failures in Aviation*, confronts the issue squarely: "Whimpington is a colossal moron of the Nth degree. If Hooper flew, then so does a dropped rock."

THE POLITICS OF POWER

The final point of contention, and by no means the least, concerns the question of powered flight. The vital question here is whether the Flungk Z-44 Mark I "Boomerang" flew under its own power. Again, there is little disagreement that the "Boomerang" was designed to be a powered aircraft; this is self-evident. After all, it had two engines. A more technically astute mind might question whether those engines were capable of sustaining the Z-44 in continuous powered flight. This becomes the key dispute among the leading authorities on Professor Flungk's body of work.

In point of fact, this debate predates the development of the Z-44, having its roots in the initial conception of the Whutley-Spitsworth "Minor Pup." Although the question of powerplant serviceability will be investigated in excrutiating detail in *Assignment: Destiny! Volume II--A Curious and Convoluted Lineage: Predecessors of the Z-44 Mk I "Boomerang,"* the serious reader will profit from a brief discussion of the Z-44 Mk I powerplants as they relate to the

Accounting officers of the Royal Flying Corps inspect some of the 1,000 Whutley-Spitsworth "Minor Pup" engines rusting away at the abandoned Snipeley Aeromechanics, Ltd. hangars at Bellthornhamchester at Bleak.

current discussion of airworthiness.

From its inception in 1914, the Whutley-Spitsworth "Minor Pup" was an aircraft engine unequalled in the annals of aviation history for the amount of controversy it was able to create. The staunchest defenders of the engines were few, but powerful, and quite frankly they were, almost to a man, in a position to profit financially from the implementation of the powerplant.

To quote von Heinkerblonker, in his epochal massif, *Whackos, Crackos, Sickos and Psychos--A Psychological Study of Fruitcakes, and the Nut-cases Who Warped Them*: "Quite often it is surprising to the layman, but socio-economic factors may command much more influence in the technological decision-making process than the actual restraints imposed upon the specifications of a particular project by the cold hard facts of physical and scientific limitation."

Herr von Heinkerblonker continues: "A peculiar case in point is the acceptance by the Royal Flying Corps early in World War I of an aircraft engine commonly known as the Whutley-Spitsworth 'Minor Pup,' an execrable hunk of brass, tin, rubber, and iron grossly incapable of lifting even its own weight. Even a schoolboy could see that this technological monstrosity was unfit for service."

Still, various socio-economic factors which were not immediately evident came into play. These factors caused this engine to be heaped upon an unsuspecting Royal Flying Corps in great numbers, leaving a huge cadre of RFC boffins scratching their heads in amazement as they vainly attempted to employ these aeronautical nightmares in useful service.

"It has been calculated that," von Heinblonker continues, "considering the number of skilled technicians involved who could otherwise have been employed in useful work, the huge amount of strategic raw materials consumed in manufacturing the engines, and the sheer frustration of anyone even remotely connected with this motor, the Whutley-Spitsworth 'Minor Pup' considerably aided the German war machine, possibly adding another six months duration to the Great War."

How could such an engine be produced, one might reasonably ask? Enter the socio-economic factors. By the astute transfer of large sums of money from the owners of the Whutley-Spitsworth Affordable Engines, Ltd. firm to various members of the House of Lords, political pressure was brought to bear on the staff of the Royal Flying Corps, making the purchase of the "Minor Pup" engine a desirable alternative to early retirement and possible loss of a government pension for the RFC officers involved. Additional influence was brought to bear by members of the House of Commons. Additionally, it seems that, as each Whutley-Spitsworth 'Minor Pup' engine required the retention of at least one ladling officer, a huge voting bloc was inadvertently created with the formation of The International Brotherhood of Sump Ladlers, or I.B.S.L. Thus both the House of Lords and the House of Commons of the British Parliament had vested interests in seeing to the further implementation of Whutley-Spitsworth Affordable Engines, Ltd. products.

Obtuse to the point of absurdity, Dr. Whimpington nonetheless established a firm baseline, if not a beachhead, in the continuing quest for a definitive answer to the question of flight controllability in regards to the Flungk "Boomerang" aircraft.

AN INTELLECTUAL STORM OF MONSTROUS PROPORTION: SERVICEABILITY CONCERNS OF THE "MINOR PUP"

This sad commentary continues to this day, due to the unfortunate choice by Professor Herrmann Flungk of the Whutley-Spitsworth "Minor Pup" engine as the powerplant for his infamous Flungk Z-44 Mark I "Boomerang." In a concise quotation from Group Commander Reichard Kolbenfente, DVM&LLC, in his insightful and revealing omnibus edition, *From the Pike of Possibility to the Bludgeon of Disaster: The Star-Crossed Careenings of the Flungk ZXPVT-1 Boomerang*, we find the following: "The clamorous clang of discord amongst current historians is reminiscent of nothing so much

as the clatter of valves and howling of bearings of a screaming Whutley-Spitsworth 'Minor Pup' at full throttle, whose frustrated ladling officer, in a fit of angst and despair, has wandered off in search of the nearest pub."

Dr. Eustis Bothomfieder in his concise treatise, *They Should Have Been Shot: Scammers, Pilferers, and Looters of the British Army Quartermaster Corps*, views the problem as follows: "The 'Minor Pup' engine was a piece of crap, and, as far as I'm concerned, Whutley and Spitsworth both should have been awarded the German Iron Cross for their service to the German war effort."

Even so august a personage in the field of Flungk triviata as Halloway Bumpsteed, Jr. has commented on numerous occasions in his volume *They Might Have Been Giants: Misadventures, Blunders, & Colossal Failures in Aviation*, on the Whutley-Spitsworth "Minor Pup" engine, most favorably comparing it to a boat anchor. In light of the above arguments, the consensus of the experts seems to be that, at the absolute very best, the Whutley-Spitsworth "Minor Pup" engines may possibly have lacked sufficient power to adequately propel the Flungk Z-44 Mark I "Boomerang" through its full envelope of flight modes. Indeed, as has been seen, there exists a strong viewpoint within the aeronautical design community that the only flight mode that a Whutley-Spitsworth "Minor Pup" equipped aircraft would be capable of would be descent. Still, on the day in question, the evidence seems incontrovertible that the Z-44 did indeed depart the earth and come to rest at a point some distance away from its starting point. Yet, was this due to the inherent characteristics of the aircraft, or, was it due solely to external environmental influences, in this case, winds in excess of forty knots? And, even with the help of the wind, did the obviously underpowered "Boomerang" nonetheless demonstrate the validity of Professor Flungk's theories sufficiently to warrant subsequent development?

Here the question stands to this day. Even in the light of this cacophonous discord, one incontrovertible fact can be agreed upon by all the participants involved. Without a doubt, there is absolutely no argument that Professor Herrmann Flungk at the very least created a non-controllable powered kite capable of carrying at least one passenger in excess of twenty feet.

WHY WAS THE BOOMERANG THE Z-44 MARK I?

Multitudes of researchers, at least three at present knowledge, have combed courthouse records, airport documents, and the papers of Prof. Flungk himself, for years. They have been searching for any hint, however slight, to the fate of the previous 43 models of the Flungk Z-44 Mk I "Boomerang" model series. In this search, they have been largely unsuccessful. Much speculation has ensued upon the choice of the desig-

nation letter "Z" in the nomenclature. In his massive treatise, von Heinkerblonker states that, ". . . the letter 'Z' obviously stood for a level of experimentation at least two orders of magnitude beyond the relatively simple and unsophisticated experimental 'X' planes of the United States government. Oddly, it is a sad fact that no previous models of the 'Boomerang' series of aircraft, the models Z-1 through Z-43, or any drawings, receipts, or documentation of any kind exist. This alone attests to the tremendous wall of secrecy that Prof. Flungk had erected around his project. The fact that the 'Z' series was further subdivided into Mark designations, as in the 'Z-44 Mark I,' stands in mute testimony to the tremendous number of revisions that the 'Boomerang' series underwent in order to create the finely honed flying machine that it ultimately became."

According to Leo Hooper, Jr., who was present during the early development phase, if in only a custodial capacity, the true facts of the case were somewhat different. To quote Mr. Hooper:

"You see, the Professor wanted to call it the Z.S. of A.M.&B.T.-1 Flungk Fleugener." Leo said.

The occasion of Hooper's statement was an interview by a reporter for the Zanesville Weekly Standard on August 14, 1958 at the Lucasville Correctional Facility. Although the reporter was trying to get information on an unrelated series of burglaries attributed to Leo, it was extremely difficult to get him to talk about anything except his days at the Zanesville School of Aviation Maintenance & Beauty Technology, which he obviously considered to be his glory days.

Continuing Mr. Hooper's quote:

"I says, 'No way, Professor, that's too much of a mouthful! You need something snazzy, like-Hey, a boomerang! That's what it is, ain't it? And you need to make them letters zippy too. Make it a Z model, for Zanesville, OK? And hey, let's make it a big number, say 12, or 14, or, hey! How about 44? Then the rubes will think you made all these other models before, and put all this work into it, and they'll give you more dough! Then we put that 'Mark I' on the end, to make it sound like it was a fancy sports car, or something."

Although the often self-serving and blatantly outrageous testimony of Leo Hooper must always be taken with a huge grain of salt, the above narrative does fit with the existing known facts.

Perhaps Halloway Bumpsteed, Jr. was overly harsh in his assessment of Prof. Flungk regarding the lineage and subsequent performance characteristics of the Z-44 Mk I "Boomerang." If so, it will be left to future generations to judge the validity of Bump-steed's final verdict regarding Prof. Flungk.

"Flungk was a shameless charlatan," he said in his epic opus, *They Might Have Been Giants: Misadventures, Blunders, & Colossal Failures in Aviation.* "The whole monstrous project was a sham and a grift from its non-existent beginning to its inglorious end."

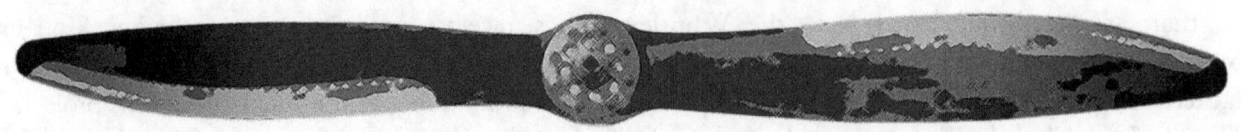

THE BOLD AIRMEN OF ZANESVILLE

We are the bold airmen of Zanesville
We are Woolsey and Hooper and Flungk
Our flights are often quite pain-filled
For we build our airplanes from junk

We are the bold airmen of Zanesville
We are Woolsey and Hooper and Flungk
With one foot in the grave
We are stalwart and brave
If the plane flies apart, we are sunk

We are the bold airmen of Zanesville
We are Woolsey and Hooper and Flungk
We are brave, we are stalwart
And we'll do our part
If only our engines will start

We are the bold airmen of Zanesville
We are Woolsey and Hooper and Flungk
We fly through the air like an anvil
Then drop to the Earth with a "Clunk!"

From "Icarus Unbound: Poems of Flight and Fury"
by Leslie Collard Hapgood

ACKNOWLEDGEMENTS

In an undertaking such as this, relating to events which occurred over a span of many years, and quite literally around the globe, the main problem for me has not been, as I initially expected it would be, the discovery and acquisition of relevant data. Rather, it has been the tedious hours of sifting and mining that raw data, searching for those golden nuggets of pertinent information among the mountains of facts, innuendo, braggadocio, misguided intentions, deliberate deception, and, quite frankly, worthless crap that, most unfortunately, composes a great deal of the documents, both governmental and private, that I and my colleagues were forced to wade through, sometimes up to our very elbows, in search of the material that composes this book

I would, of course, wish to thank those whose tireless devotion to the task at hand, and unswerving dedication to uncovering the truth behind the veil of official obfuscation, has so lightened my load these past years. No burden so heavy could be borne by myself alone, and without their help, I could never have attempted, much less completed, a project so arduous as "Assignment: Destiny!"

However, in acknowledging their contribution, three problems immediately arise. The first is that, because of the sensitive nature of much of the information contained herein, and the rather obvious trail leading back to their respective organizations, many of my contributors have wished to remain anonymous.

Secondly, because of the frankly underhanded manner in which I have by necessity been required to gather much of the intelligence contained in this work, many contributors have expressed varying degrees of reluctance in wishing to be associated with "Assignment: Destiny!", up to and including the desire that I be boiled in oil, lynched from the highest tree, or some other such nonsense. One researcher has even contributed an aviation-oriented suggestion that I take an "aeronautical intercourse upon a motivated pastry," though that is not the exact terminology that he employed. I am fairly certain that they are merely engaging in playful hyperbole, and imply no serious threat. Nonetheless, I am keeping my doors locked.

Thirdly, there sadly exists, within the aviation intelligence community, a group of researchers who question in varying degrees the findings presented in this volume, regardless of the evidence staring them baldly in the face. They know who they are, and it would be small-minded of me indeed if I were to name names.

Nonetheless, there are four people who above all else have been particularly helpful in this tremendous undertaking. To protect their reputations, I have deliberately obscured their identities.

"Ms. Amanda Floss," as my executive assistant, has proven invaluable in keeping me on track through the many grueling nights when there appeared no end in sight. Her numerous renditions of the solos from the various Wagnerian operas brought a warmth and vitality to many a long hour of reading through a pile of papers as cold as dead fish.

And, it hardly needs to be said that I would have been a mere technical neophyte adrift in a vast roiling sea of mechanical knowledge without my rudder, my stern-man, my beacon in the storm, "Dr. Quachim Z. Pipette."

As Senior Chair of the Department of Aeronautical Engineering at a prestigious university which of course must go unnamed, the good "Dr. Pipette" resolutely and

Ms. Amanda Floss. Thanks too, "Mandy," for the many massages.

concisely answered every aviation question of a technical nature which I encountered while compiling "Assignment: Destiny!" Still, I must take the credit for any factual inaccuracies which occur in this book, much as I would like to unload them on the good Doctor, who, after all, represented himself as the ultimate authority on all things aeronautical.

No profession of gratitude would be complete if I failed to mention my writing mentor of these many years, that maestro of the diphtong, the one man who molded me into the churning powerhouse of investigative reporting that I am today, Professor Phishwitter P. Wraingweld.

This is the giant of Journalism who taught me all I know about representing my ideas to the public at large. I am still working on those incompletes, Professor. Perhaps we could have lunch sometime, if only you would answer my calls.

Last, but certainly not least, is my editor, "Mr. X," a man of infinite patience. "Mr. X" thankfully works at a publishing house which is big enough to provide him with the anonymity of numbers. With over fifty editors at this prestigious house, it is hardly likely that his involvement with this book will be discovered. His vicious butchering of my prose has long ago been forgiven, as has his constant belittlement of my work, and his constant badgering of me to "get up off of my duff and write something worthwhile." I am not a man to hold a grudge. Therefore, he shall remain nameless. Isn't that right, "Mr. X?" Or, should I say, Christoff? Ha, ha, just kidding!

In addition to those listed above, there are numerous individuals who contributed to this massif. Although many cannot be named for reasons of job security, the following I can express my gratitude to with a minimum fear of retribution, as they are variously protected by academic tenure, pre-nuptial agreements,

Dr. Quachim Z. Pipette, my aeronautical mentor. C'est la vie, Quachim? Non, non, c'est la guerre! Ha ha!

Professor P. P. Wraingweld

statutes of limitations in the countries in which they reside, post-lawsuit non-disclosure agreements, or a combination thereof:

Dr. Timothy K. Valentine of RocketsRUs; all the staff at the Imperial Cricket Museum; the late Whitley Speale; Chevelle Speidelle; Captain Sir Geoffery Blunt, RFC, DMC, MBA, ETC; Dr. Egevny Putoffski of the Russian Institute of Advanced Astroplatonic Research; Commander William "Wild Bill" Gzhorwortz of VMF 301; Captain Hal Ewing for his assistance on death-ray research; Warden Jefferies and the record room trustees at the Waddingham Correctional Institute; Frau Hilda GraunSchteen, her father, Herr Tobell, and her lovely daughter Ingrid GraunSchteen, for their hospitality during my convalescence; Captain Dan Linkous and all the guys at Vultee; Dr. Bellamy Snickley of the U.S. Bureau of Census; Ralph and Dave Kelley and the Bone Lake Research team; the Flying Peckerheads of the 363rd Transport Command; Dr. Robert Fulkmore of the National Institute of Mental Health, Bethesda, Md.; Dr. Robert Jarvis of Brewster Aerospace (the Poughkeepsie division); Annabelle Cheroot and all the gang at the D.O.D. Office of Obscuration; Dr. Waldo von Heinkerblonker, M.S., PhD, L.S.M.F.T. for his kind permission to quote from his epochal tome *Whackos, Crackos, Sickos and Psychos--A Psychological Study of Fruitcakes and the Nutcases who Warped Them*; Ginger and Claire in the mail room; Capt. Tom Garrett of the Ohio State Police Air Wing; all the gang at Robert Newlon Airpark, and the West Virginia Skydivers; Dr. Manfredd von Goetzzenberger of Das FleugelWerkes; of course, the inimitable Halloway Bumpsteed, Jr. for generous access to his monumental work, *They Might Have Been Giants: Misadventures, Blunders, & Colossal Failures in Aviation*; and last, but not least, all the little tin-pot pip-squeaks at every government agency who put me on hold, transferred me to a dead line, said they would check on something and then laid the phone down for fifteen minutes, or told me that my question wasn't the concern of their department.

Truly, I couldn't have done this without you.

COMING SOON!

BE SURE TO WATCH FOR:
ASSIGNMENT DESTINY!
VOLUME II:
A CURIOUS AND CONVOLUTED LINEAGE--
PREDECESSORS OF THE Z-44 MK I "BOOMERANG"

The antics of Blakeley, Hooper, Flungk, et al, are endlessly fascinating. However, a true understanding of the phenomena that ultimately became the Flungk Z-44 Mk I "Boomerang" would not be complete without a thorough investigation of the ill-starred contraptions which directly preceded it.

Manic, or maniac, genius that Profesor Flungk was, his unconventional, or as some say, bizarre, approach to aeronautical innovation did not occur in a vacuum. Unfortunately, there were ample precedents.

In *Assignment: Destiny! Volume II*, the serious reader will be introduced to an entire new cast of characters, (men and women of vision, if often myopic), and their insidious contraptions. With the generous background of knowledge provided by the continued ruminations of Bumpsteed, von Heinkerblonker, Bothomfieder, and all of their various co-obfuscators, author J. Rutger Buck continues his merry roller coaster ride through the annals of obscure aviation. Here he encounters such aviation oddities as the Argyll ANC-13B "Water Whippet," His Majesty's Airship R-22, and the Snipeley AD-23C "Aerial Decoy," to name just a few.

After this mild departure from the storyline, *Assignment: Destiny! Volume III* rears its ugly head. Subtitled *A Girding of the Loins--Development and Deployment of the Z-44 MK I "BOOMERANG,"* this third book of the series returns the reader once again to the environs of the Zanesville School of Aviation Maintenance & Beauty Technology. From there, you will follow the transformation of the Z-44 as it wends its way through the most secretive branch of the U.S. government, the Department of Special Projects. Then the captivated reader will be entranced by the inner workings of the nascent Military-Industrial Complex, as the "Boomerang" enters full production as the YPVT-1C-EA "Indomitable Conflictor."

We think there might be a Volume IV, but we have no idea what it consists of. We're not even sure that J. Rutger Buck, the author, knows. If he ever shows up at our offices again, we'll be sure to ask him.

The Editors
Warrior Sparrow Press

www.ingramcontent.com/pod-product-compliance
Lightning Source LLC
Chambersburg PA
CBHW08081525062606
47159CB00010B/3389

9 780988 536463